PRAISE FOR THE *MANGA GUIDE* SERIES

"Highly recommended."
—CHOICE MAGAZINE ON *THE MANGA GUIDE TO DATABASES*

"The *Manga Guides* definitely h[illegible]
—SMITHSONIAN MAGAZINE

"The art is charming and the hu[illegible] and fairly painless lesson on what many consider to be a less-than-[illegible] subject."
—SCHOOL LIBRARY JOURNAL ON *THE MANGA GUIDE TO STATISTICS*

"Stimulus for the next generation of scientists."
—SCIENTIFIC COMPUTING ON *THE MANGA GUIDE TO MOLECULAR BIOLOGY*

"The series is consistently good. A great way to introduce kids to the wonder and vastness of the cosmos."
—DISCOVERY.COM

"Absolutely amazing for teaching complex ideas and theories . . . excellent primers for serious study of physics topics."
—PHYSICS TODAY ON *THE MANGA GUIDE TO PHYSICS*

"A great fit of form and subject. Recommended."
—OTAKU USA MAGAZINE ON *THE MANGA GUIDE TO PHYSICS*

"I found the cartoon approach of this book so compelling and its story so endearing that I recommend that every teacher of introductory physics, in both high school and college, consider using it."
—AMERICAN JOURNAL OF PHYSICS ON *THE MANGA GUIDE TO PHYSICS*

"This is really what a good math text should be like. Unlike the majority of books on subjects like statistics, it doesn't just present the material as a dry series of pointless-seeming formulas. It presents statistics as something *fun* and something enlightening."
—GOOD MATH, BAD MATH ON *THE MANGA GUIDE TO STATISTICS*

"A single tortured cry will escape the lips of every thirty-something biochem major who sees *The Manga Guide to Molecular Biology*: 'Why, oh why couldn't this have been written when I was in college?'"
—THE SAN FRANCISCO EXAMINER

"A lot of fun to read. The interactions between the characters are lighthearted, and the whole setting has a sort of quirkiness about it that makes you keep reading just for the joy of it."
—HACKADAY ON *THE MANGA GUIDE TO ELECTRICITY*

"The *Manga Guide to Databases* was the most enjoyable tech book I've ever read."
—RIKKI KITE, LINUX PRO MAGAZINE

"*The Manga Guide to Electricity* makes accessible a very intimidating subject, letting the reader have fun while still delivering the goods."
—GEEKDAD

"If you want to introduce a subject that kids wouldn't normally be very interested in, give it an amusing storyline and wrap it in cartoons."
—MAKE ON *THE MANGA GUIDE TO STATISTICS*

"A clever blend that makes relativity easier to think about—even if you're no Einstein."
—STARDATE, UNIVERSITY OF TEXAS, ON *THE MANGA GUIDE TO RELATIVITY*

"This book does exactly what it is supposed to: offer a fun, interesting way to learn calculus concepts that would otherwise be extremely bland to memorize."
—DAILY TECH ON *THE MANGA GUIDE TO CALCULUS*

"Scientifically solid . . . entertainingly bizarre."
—CHAD ORZEL, SCIENCEBLOGS, ON *THE MANGA GUIDE TO RELATIVITY*

"Makes it possible for a 10-year-old to develop a decent working knowledge of a subject that sends most college students running for the hills."
—SKEPTICBLOG ON *THE MANGA GUIDE TO MOLECULAR BIOLOGY*

"*The Manga Guide to the Universe* does an excellent job of addressing some of the biggest science questions out there, exploring both the history of cosmology and the main riddles that still challenge physicists today."
—ABOUT.COM

"*The Manga Guide to Calculus* is an entertaining comic with colorful characters and a fun strategy to teach its readers calculus."
—DR. DOBB'S

THE MANGA GUIDE™ TO PHYSIOLOGY

THE MANGA GUIDE™ TO
PHYSIOLOGY

ETSURO TANAKA,
KEIKO KOYAMA, AND
BECOM CO., LTD.

THE MANGA GUIDE TO PHYSIOLOGY.

The Manga Guide to Physiology is a translation of the Japanese original, *Manga de wakaru kisoseirigaku*, published by Ohmsha, Ltd. of Tokyo, Japan, © 2011 by Etsuro Tanaka, Keiko Koyama, and BeCom Co., Ltd.

This English edition is co-published by No Starch Press, Inc. and Ohmsha, Ltd.

Printed in USA
First printing

19 18 17 16 15   1 2 3 4 5 6 7 8 9

ISBN-10: 1-59327-440-8
ISBN-13: 978-1-59327-440-5

Publisher: William Pollock
Production Editor: Laurel Chun
Author: Etsuro Tanaka
Illustrator: Keiko Koyama
Producer: BeCom Co., Ltd.
Developmental Editors: Liz Chadwick, Seph Kramer, and Tyler Ortman
Translator: Arnie Rusoff
Technical Reviewers: Alisha Lacewell, Dan-Vinh Nguyen, and Kevin Seitz
Copyeditor: Fleming Editorial Services
Compositor: Laurel Chun
Proofreaders: Kate Blackham and Serena Yang
Indexer: BIM Indexing & Proofreading Services

Opener illustrations for Chapters 1, 3, 5, 6, 7, and 10 designed by Freepik.

For information on distribution, translations, or bulk sales, please contact No Starch Press, Inc. directly:
No Starch Press, Inc.
245 8th Street, San Francisco, CA 94103
phone: 415.863.9900; info@nostarch.com; http://www.nostarch.com/

Library of Congress Cataloging-in-Publication Data
Tanaka, Etsuro.
[Manga de wakaru kisoseirigaku. English]
The manga guide to physiology / by Etsuro Tanaka, Keiko Koyama, and BeCom Co., Ltd.
pages cm
Includes index.
Summary: "A guide to human physiology that combines Japanese-style manga cartoons with educational content. Topics include the circulatory system, respiratory organs, digestive system, and the brain and nervous system, as well as concepts like genes, reproduction, and the endocrine system"-- Provided by publisher.
ISBN 978-1-59327-440-5 -- ISBN 1-59327-440-8
1. Human physiology--Comic books, strips, etc. 2. Graphic novels. I. Koyama, Keiko. II. BeCom Co. III. Title.
QP34.5.T36318 2016
612--dc23
2015030911

CONTENTS

PREFACE

If you're reading this book, you may well be a student in a medical-related field, so you know how daunting the human body can be as a study subject. But once you learn a little more about it, you'll see that the human body actually has a very logical organization and that it's not as hard to learn about as you think. It always amazes me how cleverly the human body is organized.

Unfortunately, many people are reluctant to learn physiology because it seems like there is so much to cover, with so many different areas of study, that starting the subject can be intimidating. This is a shame, because once you know the basics of how the body works, it really is a fascinating subject. This book attempts to convey the magnificence of the human body in an enjoyable and easy-to-understand manner.

The Manga Guide to Physiology uses the story of Kumiko, who has previously struggled with her physiology class, to make understanding physiology fun. By getting a first-hand feel for physiology through her own physical experiences, like eating and running, Kumiko quickly develops a personal interest and begins to enjoy learning, and hopefully you will too.

If you've never studied physiology before or have found it difficult to grasp, reading through the comic sections first will give you a decent overview. Reading through both the comic and the text sections together will give you a more detailed understanding.

If this book helps you understand physiology, it will give me great pleasure as its editor.

ETSURO TANAKA
NOVEMBER 2011

PROLOGUE

WHAT DO YOU MEAN I HAVE TO TAKE PHYSIOLOGY 101?

NEXT WEEK'S STUDENT COUNCIL–SUPPORTED MARATHON WILL WIND ITS WAY THROUGH THE VAST CAMPUS.
KOUJO MEDICAL SCHOOL
Student Council–Sponsored
MARATHON
IT'S A FAMOUS EVENT IN THE SMALL UNIVERSITY TOWN AND CAUSE FOR MUCH CELEBRATION.
ADJUNCT LECTURER ROOM
MS. KARADA,
YOUR DEDICATION TO TRAINING FOR OUR SUMMER MARATHON IS ADMIRABLE, BUT CLEARLY YOU HAVE LET IT INTERFERE WITH YOUR STUDIES!
YOU'RE THE ONLY ONE IN MY CLASS WHO FAILED!
PHYSIOLOGY LECTURER *MITSURO ITANI*
SCHOOL OF NURSING, KOUJO MEDICAL SCHOOL
I...I CAN'T BELIEVE IT...
FRESHMAN NURSING STUDENT *KUMIKO KARADA*

IF YOU KEEP THIS UP, YOU'RE GOING TO WASH OUT!
YOU CAN'T BECOME A REGISTERED NURSE UNLESS YOU LEARN PHYSIOLOGY.
BU-BUT...
PROFESSOR IYAMI... PLEASE... I'LL...
THE NAME IS *ITANI*!!!!
DESPITE YOUR FOOLISHNESS, I CANNOT GIVE UP ON YOU.
MUMBLE
ESPECIALLY BECAUSE MY OWN EVALUATION SCORE WOULD SUFFER...
MUMBLE
SO, I GUESS I SHOULD GIVE YOU A SPECIAL MAKEUP EXAM.
THA-THANK... YOU.
YOUR MAKEUP EXAM WILL BE IN TEN DAYS!
MEMORIZE EVERYTHING IN MY TEN FAMOUS BOOKS!
PLOP!

HMM....
THERE'S JUST NO WAY...
I WON'T BE ABLE TO DO THIS. EVEN IN HIGH SCHOOL, I WAS ALWAYS BAD AT MEMORIZING...

HRM... MAYBE I CAN FIND SOME OTHER JOB IN HEALTH CARE.

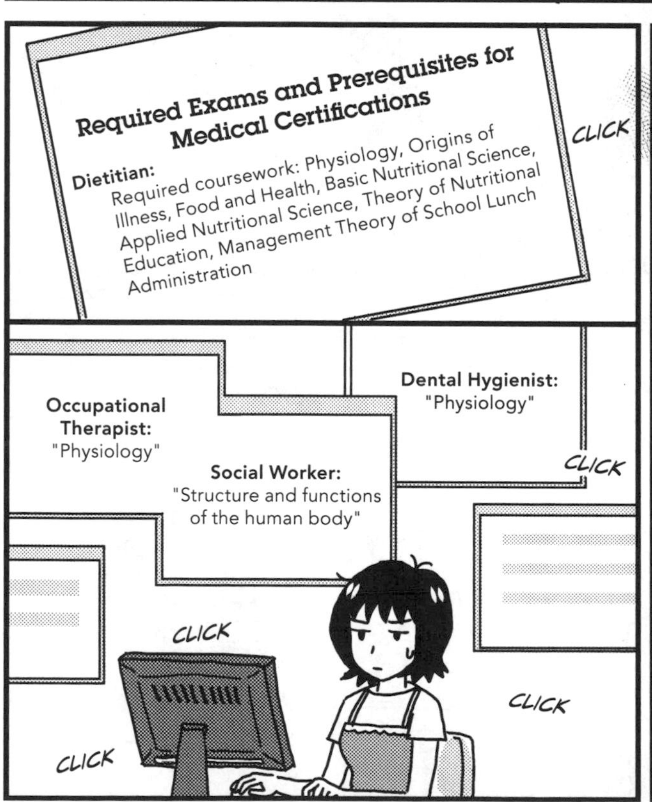
Required Exams and Prerequisites for Medical Certifications
Dietitian:
Required coursework: Physiology, Origins of Illness, Food and Health, Basic Nutritional Science, Applied Nutritional Science, Theory of Nutritional Education, Management Theory of School Lunch Administration
CLICK
Occupational Therapist:
"Physiology"
Dental Hygienist:
"Physiology"
Social Worker:
"Structure and functions of the human body"
CLICK
CLICK
CLICK
CLICK

EVERYTHING INVOLVES PHYSIOLOGY!
I'M GOING TO HAVE TO LEARN THIS NO MATTER WHAT, AREN'T I?!
BLARGH!

...BUNDLE OF HIS...
...PURKINJE FIBERS...
GRUMBLE GRUMBLE
DEPARTMENT OF SPORTS AND HEALTH SCIENCE
GRUMBLE GRUMBLE
GRUMBLE GRUMBLE
HEALTH SCIENCE LAB
CREEEAK

Mitsuro Itani
PHYSIOLOGY
for Registered Nurses
YE-OWWW!
GET BACK, ZOMBIE FIEND!
CLONK!
CRACK!
WHOA WHOA WHOA WHOAAA!
CRASH!
BANG!
PHYSIOLOGY...?
Mitsuro Itani
PHYSIOLOGY
for Registered Nurses
PHYSIOLOGY

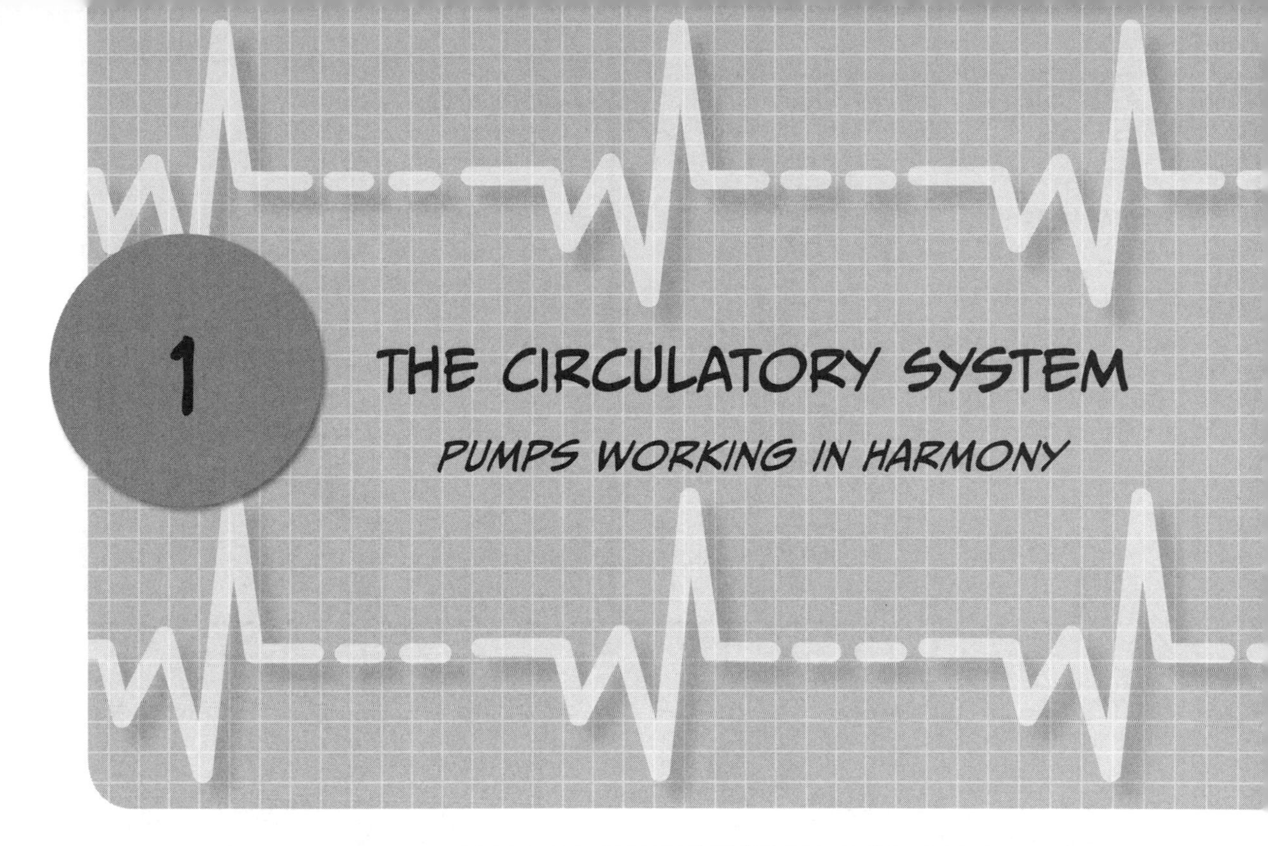

1 THE CIRCULATORY SYSTEM

PUMPS WORKING IN HARMONY

THE ELECTRICAL CONDUCTION SYSTEM OF THE HEART

SO LISTEN...
YOU SAID YOU'RE TAKING A MAKEUP EXAM...IS THAT BECAUSE YOU DISLIKE PHYSIOLOGY?

NO, I *HATE* IT...
HARUMPH
ER, ACTUALLY, I'M JUST NO GOOD AT IT!
I THINK A LOT OF PEOPLE FEEL THE SAME WAY.
I SEE...

I KNOW A LOT ABOUT THE BODY, BUT...
I'M A LITTLE FUZZY ON HOW IT ALL WORKS TOGETHER.

HMM...
DO YOU DO ANY KIND OF SPORTS?
YEAH.
I RAN LONG DISTANCE DURING JUNIOR HIGH AND HIGH SCHOOL.

INDEED...
LET'S HAVE MS. KARADA HELP US OUT WITH OUR PREPARATIONS.
RIGHT, GUYS?
TEACHING ASSISTANT *TOKO YAMADA*
TEACHING ASSISTANT *ATSURO SUZUKI*

* MORE THAN $12,000

THE NEXT MORNING

ESCAPE WILL BE DIFFICULT IF THEY'LL BE GUARDING ME THE WHOLE TIME...
ICY STARE
SHALL WE BEGIN?

AH, OKAY.
WHEN YOU SAID HELP OUT...
DID YOU JUST MEAN I'LL TAKE LESSONS FROM YOU?

YEAH, I NEED TO PRACTICE...
I'M TEACHING A REMEDIAL CLASS NEXT SEMESTER. SINCE YOUR PHYSIOLOGY GRADES WERE SO BAD, I THOUGHT YOU'D BE JUST THE RIGHT AUDIENCE.
MIFFED

I'LL COVER BASIC PHYSIOLOGY.
AND YOU CAN STUDY FOR YOUR MAKEUP EXAM!

THANK YOU VERY MUCH, BUT I ALREADY UNDERSTAND THE BASICS!

OH YEAH?
WELL THEN... EXPLAIN THE BASICS OF THE CIRCULATORY SYSTEM.

HUFF...

THE CIRCULATORY SYSTEM INCLUDES ORGANS THAT CIRCULATE THE BLOOD WITHIN THE BODY. THE ELECTRICAL CONDUCTION SYSTEM THAT STARTS THE CONTRACTION OF THE HEART TRANSMITS AN ELECTRICAL IMPULSE, WHICH IS A CONTRACTION INSTRUCTION, FROM THE SINOATRIAL NODE TO THE MYOCARDIUM CELLS OF THE VENTRICLE. THE SINOATRIAL NODE, THE ATRIO-VENTRICULAR NODE...UM, ER...
STOP!
STOP FOR A MOMENT. DO YOU UNDERSTAND WHAT'S COMING OUT OF YOUR MOUTH?
PANT PANT
BUT HE SAID THAT PHYSIOLOGY IS MEMORIZATION.

PROFESSOR ITANI SAID THAT?
NOD NOD

AHEM

THE CIRCULATORY SYSTEM INCLUDES THE HEART AND BLOOD VESSELS.
THE HEART IS A POWERFUL PUMP THAT SENDS BLOOD TO YOUR BODY THROUGH A NETWORK OF BLOOD VESSELS THAT ACT SOMETHING LIKE PIPES.
CIRCULATORY SYSTEM

YOU'RE REALLY GOING BACK TO BASICS, AREN'T YOU?

SURE! THE BLOOD'S JOB IS TO TRANSPORT OXYGEN AND NUTRIENTS...
AND IF THE BLOOD STOPS FLOWING, THE PERSON WILL DIE.
OUR HEART KEEPS THE BLOOD FLOWING, WHICH IS WHY IT'S SO VITAL TO KEEPING YOU ALIVE.
MAN, I STUDIED THE CIRCULATORY ORGANS A LOOONG TIME AGO.
SO...
THERE ARE TWO CIRCUITS OF BLOOD. ONE CIRCULATES THROUGH THE LUNGS AND THE OTHER THROUGH THE ENTIRE BODY.
THEY'RE CALLED *PULMONARY CIRCULATION* AND *SYSTEMIC CIRCULATION*, RIGHT?
YES, THAT'S CORRECT.
LET'S THINK ABOUT THAT SOME MORE. THE HEART IS DIVIDED INTO TWO CIRCUITS, THE *LEFT HEART* AND THE *RIGHT HEART*.
THE LEFT HEART CONTAINS THE LEFT ATRIUM AND LEFT VENTRICLE, AND THE RIGHT HEART CONTAINS THE RIGHT ATRIUM AND RIGHT VENTRICLE FOR A TOTAL OF FOUR CHAMBERS.
WELL, WELL...
SHE KNOWS THAT MUCH PERFECTLY.

SO...
MS. KARADA, WHY DON'T YOU TRY DRAWING THE HEART?
ER...
UH...
ALRIGHT.
SQUEAK
SQUEAK
HEE
HEE
IT'S SOMETHING LIKE THIS, I THINK. HOW DID I DO?
WELL... AT LEAST YOUR HEART IS IN THE RIGHT PLACE...
Right atrium
Left atrium
Right ventricle
Left ventricle
TA-DA!
SNICKER
SNICKER
KISS
KISS
. . . . . .
GRRRR
HMM...
IT SEEMS LIKE YOU'RE A REAL ARTIST IN THE MAKING.
BUT LET ME ADD A LITTLE DETAIL TO THIS, OKAY?

OKAY. ARTERIES AND VEINS ARE CONNECTED LIKE THIS TO THE HEART, WHICH HAS FOUR CHAMBERS AND FOUR VALVES.
THE LEFT AND RIGHT SIDES ARE ROUGHLY DIVIDED INTO THE LEFT HEART AND RIGHT HEART.
Right Heart
Left Heart
Vena cava
Pulmonary vein
Right atrium
Left atrium
Tricuspid valve
Mitral valve
Right ventricle
Left ventricle
Pulmonary valve
Aortic valve
Pulmonary artery
Aorta
BY THE WAY, THE LEFT HEART IS THE SIDE THAT PUMPS BLOOD TO THE ENTIRE BODY.
Systemic Circulation
Vena cava
Right atrium
Left ventricle
Entire body
Aorta
SYSTEMIC CIRCULATION CIRCULATES THE BLOOD FROM THE LEFT VENTRICLE THROUGH THE ENTIRE BODY TO THE RIGHT ATRIUM.
PULMONARY CIRCULATION CIRCULATES IT FROM THE RIGHT VENTRICLE THROUGH THE LUNGS TO THE LEFT ATRIUM.
Pulmonary Circulation
Pulmonary vein
Left atrium
Right ventricle
Lungs
Pulmonary artery
THE LEFT HEART AND RIGHT HEART ARE "PUMPS," AND SYSTEMIC CIRCULATION AND PULMONARY CIRCULATION SEND BLOOD THROUGH A SERIES OF "PIPES," RIGHT?
MAYBE IT'S NOT SO BASIC, AFTER ALL.
WELL, LET'S MOVE ON.
YOU KNOW THAT THE HEART CONTRACTS AND EXPANDS WITH A WELL-REGULATED RHYTHM WHEN THE MUSCLE CALLED THE *MYOCARDIUM* RECEIVES ELECTRICAL STIMULI.

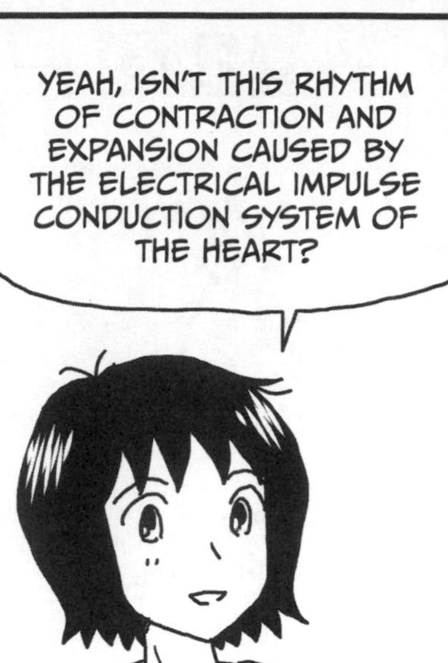
YEAH, ISN'T THIS RHYTHM OF CONTRACTION AND EXPANSION CAUSED BY THE ELECTRICAL IMPULSE CONDUCTION SYSTEM OF THE HEART?
THAT'S RIGHT!

SO...
SOCC
10
...YOU COULD CONSIDER THE IMPULSE CONDUCTION SYSTEM SOMETHING LIKE A SOCCER TEAM!
WHAT ARE YOU TALKING ABOUT...?

THE PERSON IN CHARGE OF THE ELECTRICAL IMPULSES, WHICH ARE THE SOURCE OF THE RHYTHM OF THE HEART, IS THE COACH.
Coach
Sinoatrial node

Coach
Sinoatrial node
Atrioventricular node
Team captain
Myocardium
Players
THE IMPULSES ARE TRANSMITTED TO THE CAPTAIN...
AND THEN ARE TRANSMITTED TO THE PLAYERS.

THERE IS AN EXTREMELY HIGH DEGREE OF COORDINATION BETWEEN THE CAPTAIN AND TEAM MEMBERS.
THE CAPTAIN IS LINKED TO EACH AND EVERY TEAM MEMBER BY POWERFUL BONDS.

ALSO, THE COACH IS LISTENING FOR INSTRUCTIONS FROM THE TEAM OWNER.
Owner
Brain
Sinoatrial node
¡AY-AY-AY!
IF WE SHOW THIS IN A DIAGRAM OF THE IMPULSE CONDUCTION SYSTEM...
HUH?
THE OWNER IS THE *BRAIN*. THE COACH IS THE *SINOATRIAL NODE*. THE CAPTAIN IS THE *ATRIOVENTRICULAR NODE*. THE INSTRUCTIONS FROM THE CAPTAIN ARE TRANSMITTED THROUGH THE *BUNDLE OF HIS*, *LEFT BUNDLE BRANCH* OR *RIGHT BUNDLE BRANCH*, AND *PURKINJE FIBERS*. AND THE TEAM MEMBERS ARE THE *MYOCARDIUM*.
Brain
Sinoatrial Node
Atrio-ventricular Node
Myocardium
Bundle of His
Left bundle branch
Right bundle branch
Purkinje fibers
GET OUT AND WIN, BOYS!
TODAY'S THE DAY!
KEEP UP THE PRESSURE!
LET'S DO THIS!
THE BRAIN INITIATES AN ELECTRICAL IMPULSE IN THE SINOATRIAL NODE, WHICH THEN TRANSMITS THE IMPULSE TO THE ATRIOVENTRICULAR NODE, AND THEN FINALLY TO THE MYOCARDIAL CELLS.

HEART MOVEMENTS AND WAVEFORMS

LET'S CONSIDER THE RELATIONSHIP BETWEEN THIS WAVE AND THE ELECTRICAL STIMULATION OF THE HEART.

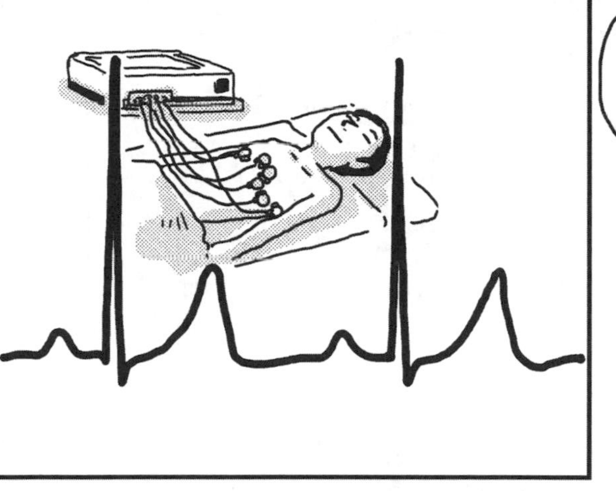

I CERTAINLY REMEMBER MY FIRST ELECTROCARDIOGRAM.

FIRST, STIMULI ARE TRANSMITTED FROM THE SINOATRIAL NODE TO THE ATRIUM, CAUSING THE LEFT AND RIGHT ATRIA TO CONTRACT.
Coach
Sinoatrial node

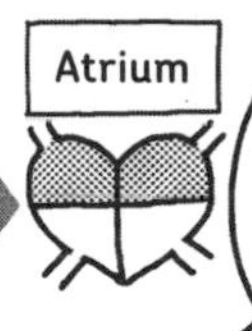
Atrium
THAT MADE THE P WAVE, RIGHT? IT'S THE LITTLEST BUMP.

THAT'S RIGHT. AND THE CONTRACTION OF THE ATRIA SENDS THE BLOOD THAT IS IN THE ATRIA TO THE VENTRICLES.

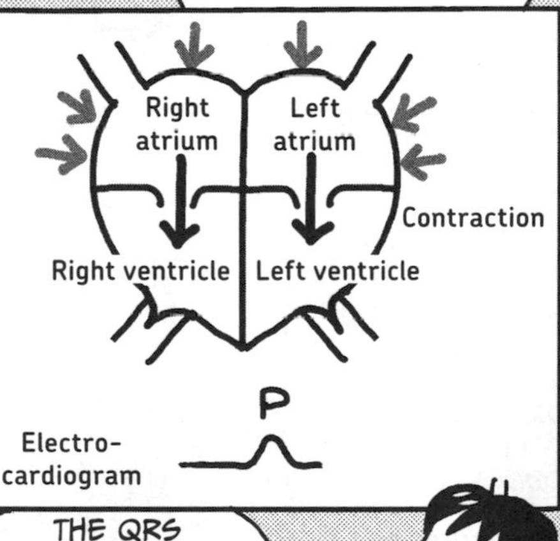
Right atrium
Left atrium
Contraction
Right ventricle
Left ventricle
P
Electrocardiogram
THE QRS COMPLEX IS TRANSMITTED TO THE VENTRICLES NEXT, ISN'T IT?

INSTRUCTIONS FROM THE ATRIOVENTRICULAR NODE ARE THEN TRANSMITTED—THROUGH THE BUNDLE OF HIS, LEFT BUNDLE BRANCH, RIGHT BUNDLE BRANCH, AND PURKINJE FIBERS—TO THE MYOCARDIUM, STIMULATING THE LEFT AND RIGHT VENTRICLES.
THIS IS THE QRS COMPLEX.
Right atrium
Left atrium
Right ventricle
Left ventricle
Contraction
R
Q
S
Electrocardiogram

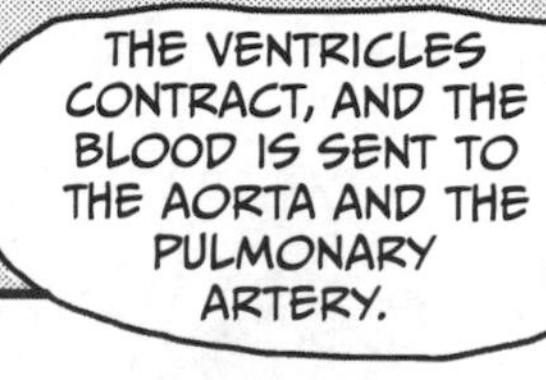
THE VENTRICLES CONTRACT, AND THE BLOOD IS SENT TO THE AORTA AND THE PULMONARY ARTERY.
RIGHT.

AND FINALLY, THE T WAVE...
THE STIMULATION OF THE VENTRICLES ENDS HERE AND THE VENTRICLES RELAX.
Right atrium
Left atrium
Right ventricle
Left ventricle
Ventricles relax
Valves close
Electro-cardiogram
T
THAT SHOULD GIVE YOU A PRETTY GOOD IDEA OF WHAT'S GOING ON!
R
P
Q
S
T
GOT IT!
BY THE WAY, DO YOU KNOW WHAT'S HAPPENING WHEN YOU HEAR YOUR HEARTBEAT?
ISN'T THAT WHEN THE VALVES CLOSE?
THAT'S RIGHT.
EACH VALVE INSIDE THE HEART MAKES A SOUND WHEN IT CLOSES, JUST LIKE A CASTANET!
THAT'S ALL FOR TODAY.
REMEMBER, IF YOU WANT TO HELP PATIENTS AND PUT YOUR KNOWLEDGE OF PHYSIOLOGY TO USE, IT WILL TAKE MORE THAN JUST MEMORIZATION. YOU ALSO HAVE TO SEE THE BIGGER PICTURE AND UNDERSTAND HOW EACH PART RELATES TO EVERYTHING ELSE!

I THINK I GOT ALL THAT....

HEY KUMIKO!
DID YOU HEAR?
YOU'RE OUR DEPARTMENT'S REPRESENTATIVE FOR THE MARATHON RACE.

OBVIOUS PICK, RIGHT?
YOU'RE THE ONLY PERSON WITH TRACK AND FIELD EXPERIENCE.

. . . .

WHAT?!
COME ON, I'VE GOT TO STUDY FOR MY EXAM!

EVEN MORE ABOUT THE CIRCULATORY SYSTEM!

The circulatory system consists of the organs that circulate blood, lymph, and other fluids throughout the body. The heart, blood vessels, and lymph nodes transport oxygen, nutrients, hormones, and the like to tissues within the body while at the same time gathering waste products from various parts of the body.

Let's learn more about how the circulatory system works.

ELECTRICAL ACTIVITY IN THE HEART

The muscle that forms the walls of the heart contracts when it receives electrical stimuli. The impulse conduction system, shown in Figure 1-1, causes this contraction to occur.

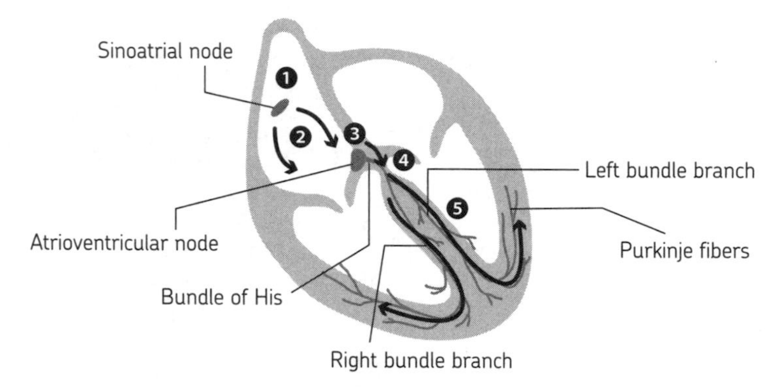

Figure 1-1: Flow of the impulse conduction system

Stimuli triggered from the *sinoatrial node* ❶ spread like waves through the entire atria, causing the atria to contract. The stimuli reach the *atrioventricular node* ❷, which is located between the left and right atria, and are transmitted to the *bundle of His* ❸. The bundle of His is divided into two branches ❹, the *left bundle branch* and *right bundle branch*. The left bundle branch and right bundle branch are further divided into numerous finer branches in the left and right ventricles, respectively. These finer branches are the *Purkinje fibers* ❺. The impulse conduction system resides in specialized muscle tissue called *cardiac muscle*, or *myocardium*.

The sinoatrial node automatically generates the stimuli, right?

That's right. It generates 60 to 80 stimuli per minute, even if it receives no instructions from the central nervous system. In other words, the sinoatrial node generates the normal heart rate and thereby acts as the heart's natural pacemaker.

Stimuli are also generated by other cardiac fibers, such as those in the atrioventricular node. However, the sinoatrial node normally controls the heart rate because it discharges stimuli faster than does any other part of the heart. If the sinoatrial node malfunctions, the atrioventricular node becomes the pacemaker in its absence. But since the atrioventricular node generates stimuli at a slower pace, the heart rate decreases when stimulated by that node.

HOW AN ELECTROCARDIOGRAM WORKS

An electrocardiogram is a visualization of the electrical stimuli transmitted to the entire myocardium from the impulse conduction system. Normally, six electrodes are attached to the chest, and a total of four electrodes are attached to both wrists and both ankles. (Electrodes connected to both wrists and one ankle take the electrocardiogram; the right ankle is attached to a neutral, or *ground*, lead for grounding the circuit.) This lets us measure the electrocardiogram using 12 leads (see Figure 1-2).

Limb Leads	**Chest Leads**
Electrodes recording at three locations (right wrist, left wrist, and left ankle), plus one grounding lead (right ankle)	Electrodes at six locations surrounding the heart
Help doctors and nurses get a "view" of the heart in the vertical plane (a coronal plane)	Help doctors and nurses get a "view" of the heart in the horizontal plane (a transverse plane)
I, aV_R, aV_L, II, aV_F, III	V1 V2 V3 V4 V5 V6

The six leads named I, II, III, $_aV_R$, $_aV_L$, and $_aV_F$ are called limb leads, and the six leads named V_1 to V_6 are called chest leads.

Figure 1-2: A 12-lead electrocardiogram

Why are 12 leads required? That seems like a lot.

Consider the leads to be something like cameras viewing the heart from the sites where the electrodes are attached. With that many camera angles on the scene, you've got the complete, 3D picture, and not much can be missed.

If there are well-regulated contractions of the heart, normal waveforms will appear in a continuous loop. However, if there is an abnormality in the myocardium or impulse conduction system, various changes will appear in the corresponding waveform of the electrocardiogram. For example, if there are *arrhythmias*—heart contractions with unusual timing—irregular waveforms will appear. Other types of arrhythmia are tachycardia, a heart rate that is too high, and bradycardia, one that is too low.

So approximately how much blood do you think is sent to the aorta each time the heart contracts?

Hmm . . . about a soda can's worth?

Whoa . . . wait a minute. The heart is about the size of a fist. There's no way it holds 350 milliliters. The so-called *stroke volume* of the heart is approximately 70 milliliters. That's about the size of a small bottle of perfume or pudding cup.

We can calculate the cardiac output per minute as follows:

Cardiac Output (mL/min) = Stroke Volume (mL/beat) × Heart Rate (beats/min)

DID YOU KNOW?

The heart rate of an infant is faster than that of an adult; it slows as the child ages. Most adults have a resting heart rate of about 60–80 beats per minute. An elderly person tends to have a slightly slower heart rate than a young or middle-aged adult.

Since the circulating blood volume in the human body is approximately 5 liters, all the blood circulates through the entire body in approximately 1 minute.

HOW THE NERVOUS SYSTEM AFFECTS THE CIRCULATORY SYSTEM

Your heart rate increases when you're surprised, speaking in front of an audience, playing sports, or in other stressful situations. This increase is caused by the activity of your autonomic nervous system (see page 138). If more blood flow is required due to stress or exertion, the sympathetic nervous system is excited, the sinoatrial node is stimulated, and your heart rate increases. On the other hand, when you relax, your parasympathetic nervous system reduces your heart rate.

But aren't the stimuli from the sinoatrial node automatically generated without receiving any instructions from the brain?

That's a good question! The sinoatrial node can certainly generate stimuli automatically, but the frequency of those stimuli is regulated by the *autonomic nervous system*.

The autonomic nervous system controls physiological responses ranging from blood pressure and heart rate to dilation of the pupils of the eyes. There are two branches of the autonomic nervous system: the *sympathetic branch* (which generates the "fight or flight" response) and the *parasympathetic branch* (which generates the "rest and digest" response).

The sympathetic nervous system is responsible for increasing the heart rate and causing blood vessel vasoconstriction (decreased diameter of blood vessels), both of which contribute to an increase in blood pressure. Conversely, the parasympathetic system is responsible for decreasing the heart rate, and the activation of the parasympathetic system leads to a decrease in blood pressure.

THE CORONARY ARTERIES

But before we begin talking about blood circulation, we should learn how the heart itself acquires oxygen and nutrients. Do you know which blood vessels send oxygen and nutrients to the myocardium?

The coronary arteries?

That's right. The *coronary arteries* are called that because they encircle the heart in a crown shape. Just think about a coronation ceremony for a new queen, where she gets her crown.

The coronary arteries are roughly divided into the right coronary artery and left coronary artery (Figure 1-3). The smaller branches of the coronary arteries penetrate the surface of the cardiac muscle mass and thus serve as the primary sources of oxygen and nutrients for the myocardium.

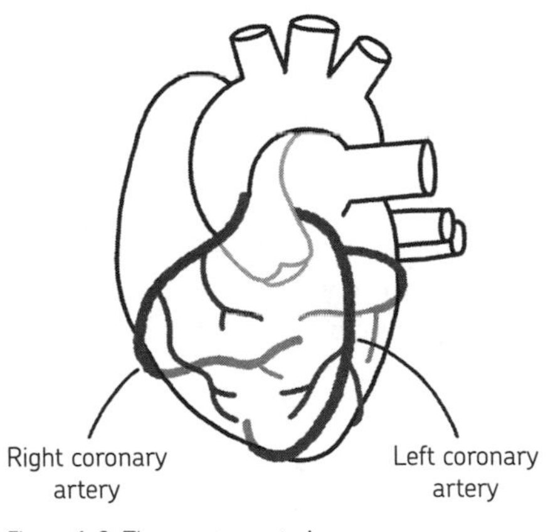

Figure 1-3: The coronary arteries

You'd think the heart would be able to get all the oxygen and nutrients it needs from the blood it is constantly pumping through its chambers. But actually, it can absorb only a minuscule amount of oxygen and nutrients that way, so the coronary arteries are needed to deliver blood deep into the muscle tissue of the heart.

The arteries of most internal organs branch and reconnect (*anastomose*). Therefore, even if a blood vessel is blocked at one location, the blood will flow along another route. However, the coronary arteries surrounding the heart are called end arteries since they are structured with no anastomoses between arterial branches (Figure 1-4). Therefore, if there is a blockage somewhere, blood will cease flowing beyond that point, causing a heart attack.

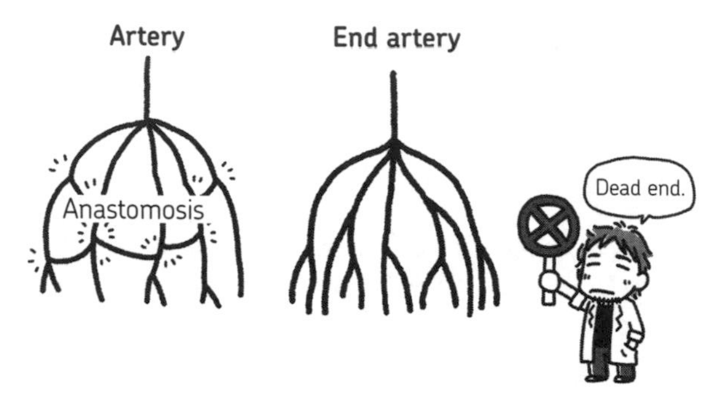

Figure 1-4: End arteries branch with no points of anastomosis.

DID YOU KNOW?

The coronary arteries aren't the only end arteries. Others are found in the brain. A blockage (or vascular occlusion) in these end arteries in the brain is very serious. A complete blockage will cause a stroke.

BLOOD CIRCULATION

We learned that there are two circuits for blood circulation: pulmonary circulation and systemic circulation. Do you think you can explain them properly?

Pulmonary circulation circulates from the right ventricle and through the lungs to capture oxygen before returning to the left atrium, and *systemic circulation* circulates from the left ventricle and through the entire body to send oxygen and nutrients to the body before returning to the right atrium.

That's exactly right! The pulmonary circulation and systemic circulation flows are depicted in a rough diagram in Figure 1-5. Since this is basic information needed for studying each of the internal organs later, make sure you understand this entire drawing.

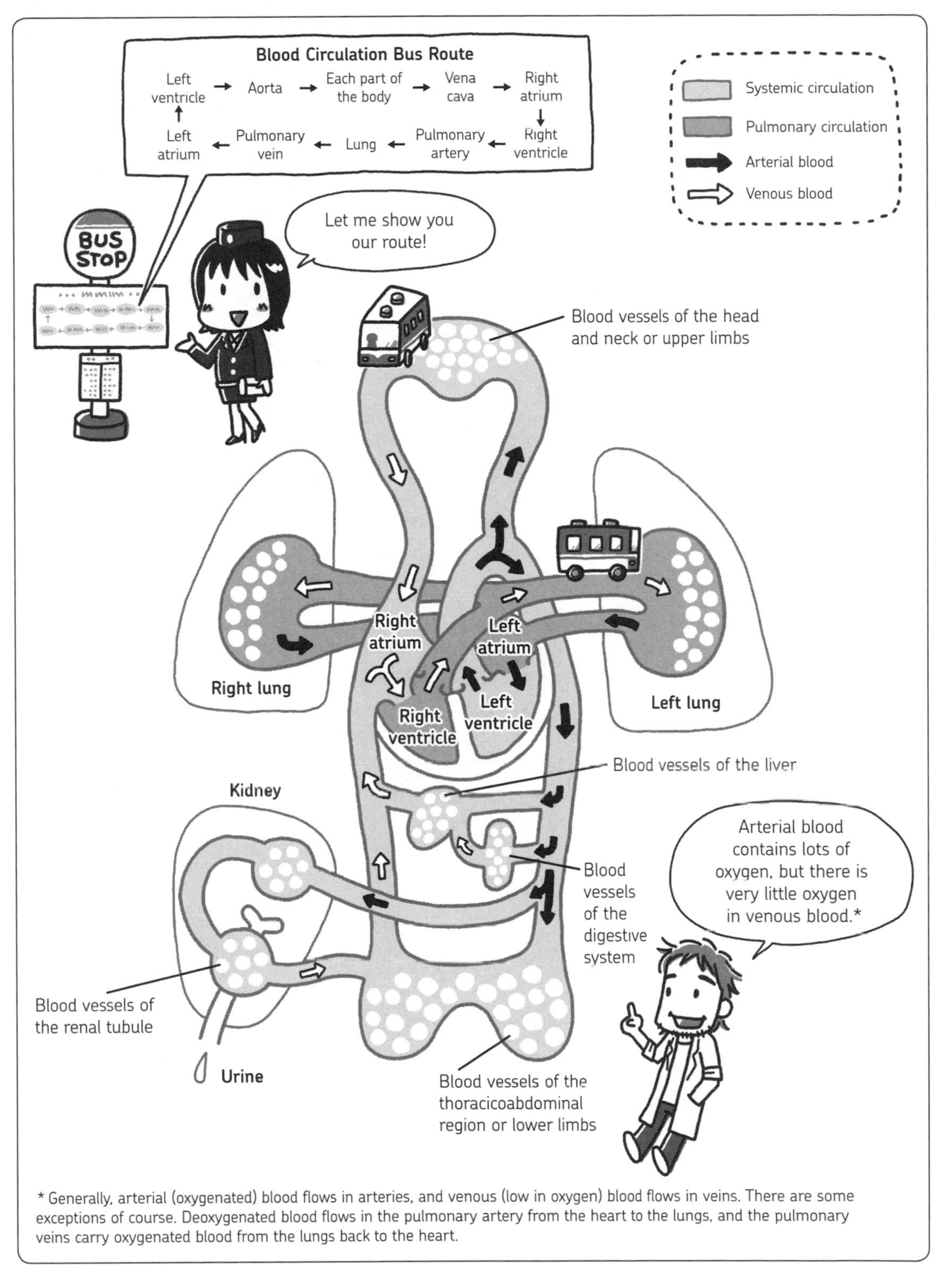

* Generally, arterial (oxygenated) blood flows in arteries, and venous (low in oxygen) blood flows in veins. There are some exceptions of course. Deoxygenated blood flows in the pulmonary artery from the heart to the lungs, and the pulmonary veins carry oxygenated blood from the lungs back to the heart.

Figure 1-5: Blood circulation

We ought to also touch on arteries and veins here. Remember that *arteries* are blood vessels carrying blood away from the heart and *veins* are blood vessels returning blood to the heart via the capillaries.

Because arteries receive blood that is pushed out of the heart under great pressure, the blood vessel walls are thick, and their elasticity and internal pressure are both high. Veins have thin blood vessel walls with valves at various locations to prevent blood from flowing backward. The internal pressure is low, and blood flow is assisted by surrounding muscles. Some veins run just below the skin. These are called superficial veins. Blood is often drawn from the median cubital vein on the inside of the elbow. This is also a superficial vein.

Although arteries often run deep within the body, they also pass through places where it is easy to take a pulse (see Figure 1-6).

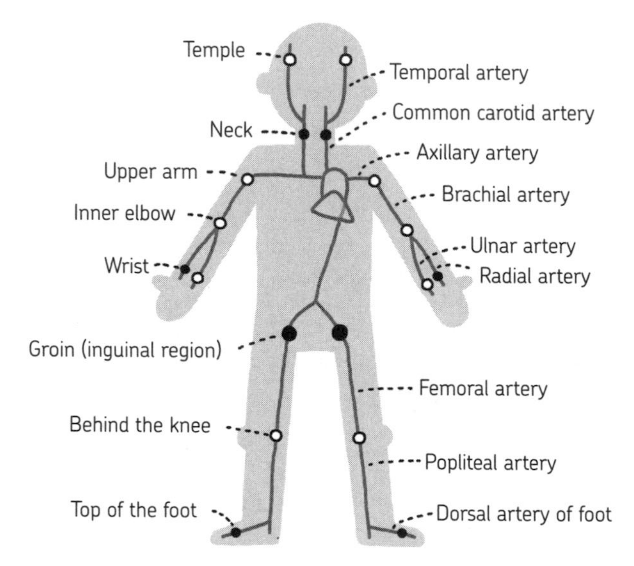

Figure 1-6: Locations for taking a pulse

These are arteries that run through locations that are relatively shallow, such as the wrist, aren't they?

That's right. In a medical clinic, your pulse is often taken using the radial artery of your wrist or the carotid artery of your neck.

BLOOD PRESSURE

Blood pressure is the internal pressure inside blood vessels, but the term is usually used to mean the pressure in large arteries near the heart, such as in the upper arm. What are some factors that determine blood pressure?

Factors? Well, er, age and diet and . . .

Yes, blood pressure certainly tends to increase as a person becomes middle aged and older, but let's consider physiological factors here.

Three factors that determine blood pressure are the girth of the blood vessels, the circulating blood volume, and the contractile force of the heart, or cardiac contractile force (see Figure 1-7). For example, if the circulating blood volume (the total volume of blood in the arteries) and the cardiac contractile force are fixed, then blood pressure will increase if the blood vessels are smaller. Also, the blood pressure will drop if the blood volume decreases because of a hemorrhage or if the contractile pressure of the heart decreases because of a heart attack.

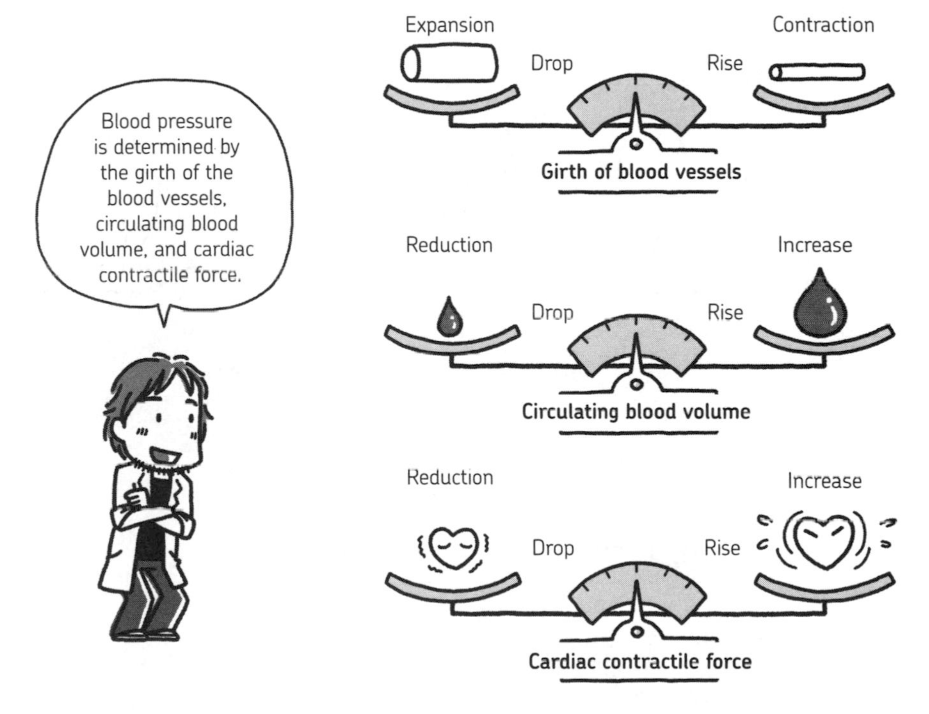

Figure 1-7: Factors that determine blood pressure

MEASURING BLOOD PRESSURE

You've studied the principles and techniques of blood pressure measurement, haven't you?

Sure, I did that in basic nursing.

Blood pressure varies like a wave, getting higher when the ventricles contract and lower when they relax. The maximum pressure is called the *systolic pressure*, and the minimum is called the *diastolic pressure*.

You inflate the cuff that's wrapped around the upper arm to restrict the blood flow. Then you release the air in the cuff a little at a time while you listen to the artery through a stethoscope. When you start to hear a tapping sound (called Korotkoff sounds), that is the systolic pressure. You continue to release air, and when you no longer can hear any sound, that is the diastolic pressure. The cuff's pressure readings at these two points give you the patient's blood pressure (see Figure 1-8).

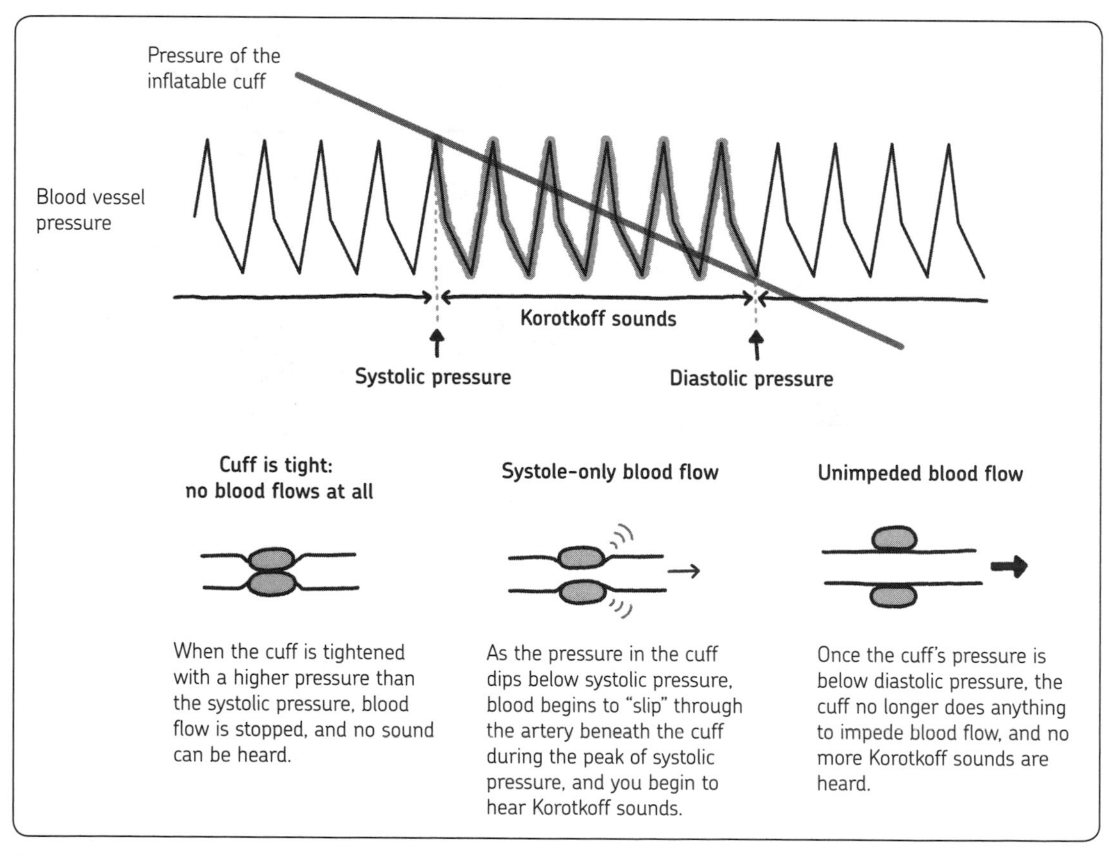

Figure 1-8: Blood pressure measurement by the auscultatory method

DID YOU KNOW?

Blood pressure units are represented by mm Hg (millimeters of mercury). The number of mm Hg indicates the number of millimeters that mercury would be pressed upward in a tube by the pressure.

THE LYMPHATIC SYSTEM

The last part of the circulatory system is the *lymphatic system*, which recovers bodily fluids that seep into tissues from capillaries and returns them to the heart. It also supports the immune system. In this way, the lymphatic organs can be said to reside in both the circulatory system and the immune system. In peripheral tissue, interstitial fluid is exchanged between capillaries and tissue, but some of the interstitial fluid is collected in the lymphatic vessels. The bodily fluid in the lymphatic vessels is called *lymph*. The lymph flow rate is approximately 2 to 3 liters per day.

The lymphatic vessels start from lymphatic capillaries, which gradually come together to form larger lymph vessels. After passing through many lymph nodes along the way, they finally enter the left and right venous angles, which are confluence points of the subclavian veins and internal jugular veins (see Figure 1-9). Valves are attached to the interior of the lymphatic vessels to prevent the lymph flow from reversing direction.

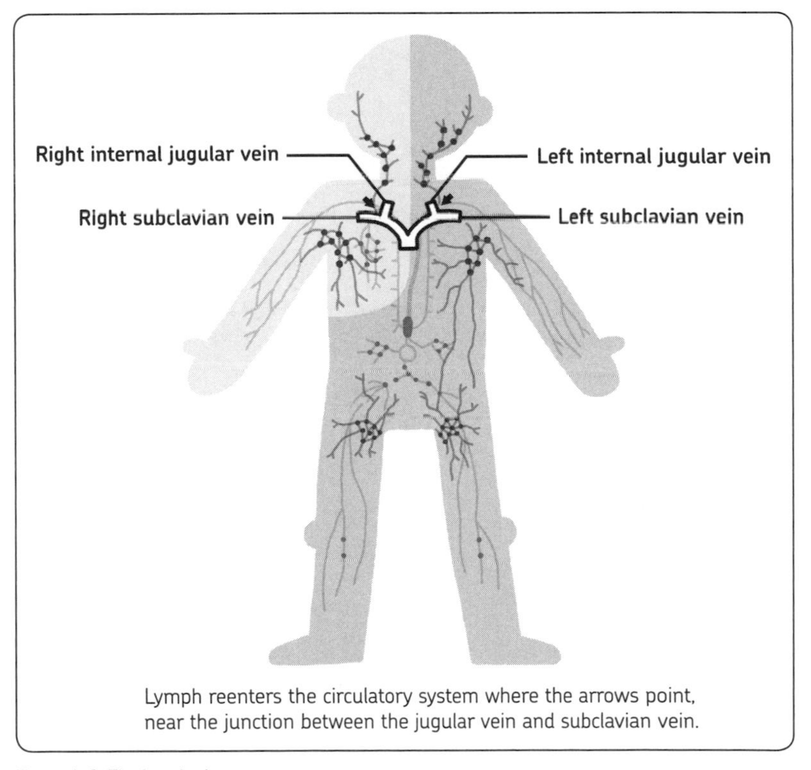

Figure 1-9: The lymphatic system

The lymphatic vessels are not symmetrical on the left and right sides of the body.

Good eye! Notice the light and dark shading in Figure 1-9. The right lymphatic trunk, in which the lymphatic vessels from the upper right half of the body are collected together, enters into the right venous angle. The collected lymphatic vessels from the remaining upper left half of the body and the entire lower half of the body enter into the left venous angle.

DID YOU KNOW?

Cancer that starts in the lymph nodes is called lymphoma. More often, cancer starts somewhere else and then spreads to lymph nodes. When cancer spreads or metastasizes, it often is found in the lymph nodes.

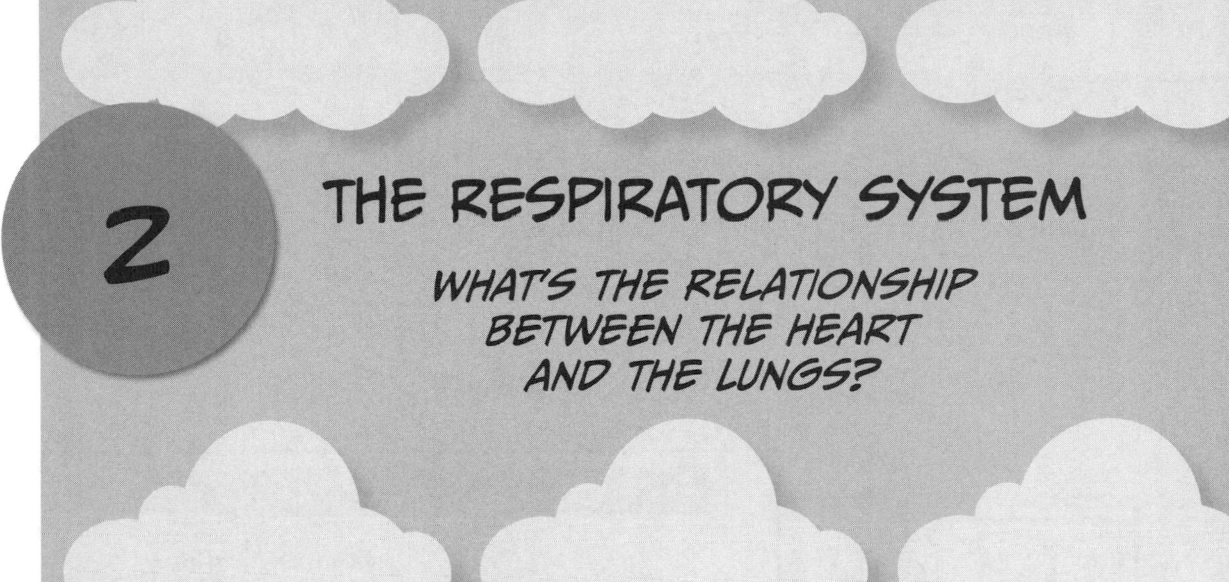

RESPIRATION'S JOB

ONE YEAR OF FREE DESSERTS AT THE KOUJO SIDEWALK CAFÉ!
WE KNOW YOU CAN DO IT... GO, KUMIKO!
ひーーっ
EEP!
9 AM
EXAM
10 AM
MARATHON RACE
ONE HOUR LATER...
GEH
GEH
ぜぜ
ぜ
ARGH!
IT'S THE SAME DAY AS THE MAKEUP EXAM? IT'S IMPOSSIBLE!
I MEAN, COME ON!
I'VE GOT TO DROP OUT!!!
はあっ
はあっ
MUTTERING TO YOURSELF WHILE RUNNING?
YOU SURE ARE A TOUGH COOKIE, MS. KARADA.

WHAT?!
PROFESSOR KAISEI?
HUFF HUFF
DON'T STARTLE ME LIKE THAT! YOU ALMOST GAVE ME A HEART ATTACK.
はぁ
はぁ
HUFF
WHAT ARE YOU DOING...
HUFF
HUFF
THIS EARLY IN THE MORNING?
HUFF
HUFF
AS YOU CAN SEE, I'M COLLECTING INSECTS.
AS MUCH AS I'M INTRIGUED BY THE AMAZING WAY THE HUMAN BODY WORKS, I ALSO THINK INSECTS ARE QUITE INTERESTING AND MYSTERIOUS.
BUT I'M ONLY AN AMATEUR ENTOMOLOGIST.
HUFF HUFF
HUFF
I SEE...
HUFF HUFF
YOU'D THINK A PROFESSOR OF SPORTS SCIENCE MIGHT HAVE A MORE ATHLETIC PASTIME!
AT ANY RATE, IT SEEMS LIKE YOU ARE HAVING TROUBLE BREATHING.
はぁ
RELAX...
BECOME CONSCIOUS OF YOUR DIAPHRAGM.
FOCUS ON THAT MUSCLE AND TAKE SLOW, DEEP BREATHS.
WHEEZE
UM, OKAY.
SUUU—
—HAHHH

THAT WORKED!

I'VE ENTERED THIS YEAR'S MARATHON RACE.
DO YOU KNOW OF ANY GOOD SHORT-TERM TRAINING METHODS?
HMM, LET'S SEE NOW...

SHORT-TERM, EH?

LET'S TALK ABOUT THE RESPIRATORY ORGANS A LITTLE.
EVENTUALLY, EVERYTHING LEADS BACK TO PHYSIOLOGY, DOESN'T IT?
YESTERDAY, IT WAS THE HEART... TODAY, IT'S RESPIRATION?

I'M SURE YOU KNOW THE ROLE THAT RESPIRATION PLAYS IN CREATING ENERGY SO YOU CAN MOVE YOUR BODY AND MAINTAIN YOUR BODY TEMPERATURE.
BOTH ARE INDISPENSABLE WHEN IT COMES TO KEEPING YOUR BODY ALIVE.

SURE. OUR BODIES MAKE ENERGY BY BREAKING DOWN FOOD AND OXYGEN. THIS IS OUR METABOLISM IN ACTION! THE WASTE OF THIS PROCESS IS THE CARBON DIOXIDE (CO_2) THAT WE EXHALE.

Automobile
Gasoline, O_2
Exhaust gas (main component is CO_2)
You can compare our metabolism to the combustion that happens in the engine of a car.
Gasoline and O_2 combust in an engine to release kinetic (or mechanical) energy.
Person

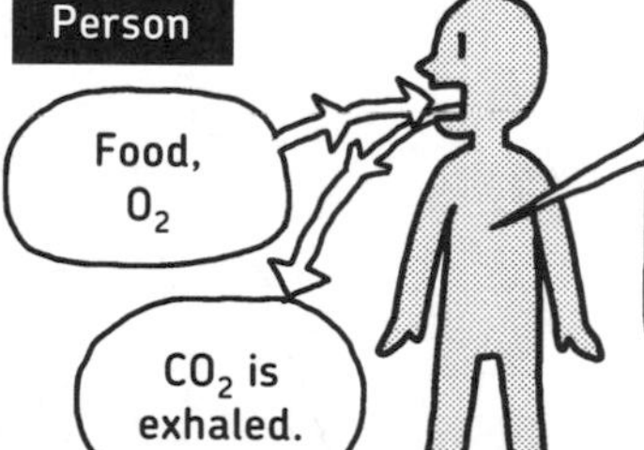
Food, O_2
CO_2 is exhaled.
Food and O_2 combust in our bodies to release chemical energy.
Feel the "burn"!

YOU'VE GOT IT JUST RIGHT. O_2 IS USED IN BOTH THE CAR AND OUR BODIES!

HOW VENTILATION WORKS
SU-HAH
はす
BY THE WAY, DO YOU REMEMBER THAT IN YESTERDAY'S LECTURE, WE LEARNED THAT BLOOD, AMONG ITS OTHER JOBS, TRANSPORTS OXYGEN?
WELL...
IT'S LIKE THIS.
Capillary
Alveolus
O_2
CO_2
SO WHAT DOES THE BLOOD DO WHILE RECEIVING OXYGEN IN THE LUNGS?
THE BLOOD RECEIVES OXYGEN FROM PULMONARY ALVEOLI.
THAT'S RIGHT.
AND AT THE SAME TIME, THE ALVEOLI RECEIVE CARBON DIOXIDE FROM THE BLOOD.
SO...
THE EXCHANGE OF OXYGEN AND CARBON DIOXIDE THAT OCCURS IN THE ALVEOLI IS CALLED *GAS EXCHANGE*.
THE ATMOSPHERE HAS SUDDENLY BECOME VERY PROFESSORIAL...
UH-OH
BLAH BLAH BLAH...
LET'S WALK AND TALK WHILE YOU COOL DOWN.

UM, PROFESSOR...
I ASKED ABOUT SOMETHING *EFFECTIVE* FOR THE MARATHON?
OH, RIGHT...
WELL, YOU UNDERSTAND HOW AIR MOVES INTO THE LUNGS, RIGHT?
YUP. THE LUNGS INFLATE AND COLLAPSE.

COULD YOU SAY IT A LITTLE MORE PRECISELY?
OF COURSE.
WELL, UM...
THE LUNGS, WHICH ARE LOCATED IN THE THORACIC CAVITY FORMED BY THE DIAPHRAGM AND CHEST WALLS, INFLATE AND COLLAPSE DUE TO PRESSURE CHANGES IN THAT THORACIC CAVITY.

GOOD... THAT'S RIGHT.
THE LUNGS DON'T INFLATE ON THEIR OWN. THEY NEED THE HELP OF YOUR MUSCLES.

ふふふ
HEH HEH HEH
MY LUNGS INFLATE BECAUSE OF THE WICKED-STRONG ABDOMINAL MUSCLES I'M DEVELOPING.
う
UH. JUST DON'T TELL ITAMI THAT ON YOUR MAKEUP EXAM.

SCRATCH
SCRATCH
SCRATCH
THERE WE GO.
HERE'S A SIMPLE DIAGRAM OF THE LUNGS AND THORACIC CAVITY.
Trachea
Lungs
Thoracic cavity
Diaphragm
MS. KARADA, YOU SAID THAT THE PRESSURE IN THE THORACIC CAVITY CHANGES, RIGHT?
EXPLAIN HOW THAT WORKS USING THIS DIAGRAM!
THE LUNGS EXPAND AS THE DIAPHRAGM MOVES DOWNWARD...
RIGHT.
AS THE DIAPHRAGM CONTRACTS, IT FLATTENS AND INCREASES THE VOLUME OF THE CHEST. WITH THIS CHANGE IN VOLUME, THE PRESSURE INSIDE THE CHEST DROPS, AND AIR RUSHES IN. THE REVERSE HAPPENS WHEN YOU EXHALE.

AND BTW...
BEEF DIAPHRAGM IS DELICIOUS.
IT'S CALLED HARAMI WHEN WE ORDER YAKINIKU GRILLED MEAT!
DROOL...
じゅるー
YUM...
I LOVE HARAMI...
SIZZLE
ジュ〜
ANYWAY, MOVING ON!
RESPIRATION THAT USES THE DIAPHRAGM IS CALLED ABDOMINAL BREATHING.
HUH?

NOW THAT YOU MENTION IT, I'VE OFTEN SEEN TIPS ABOUT ABDOMINAL BREATHING TECHNIQUES IN DIET AND EXERCISE MAGAZINES.
FITNESS Exercise!
WHY WOULD IT BE IMPORTANT TO BREATHE WITH THE DIAPHRAGM?
ABDOMINAL BREATHING IS OFTEN PRACTICED IN TAI CHI OR YOGA, AS WELL AS BY SINGERS AND THOSE WHO PLAY WIND INSTRUMENTS LIKE THE TRUMPET. IT ALLOWS FOR DEEPER BREATHS.
TAKING SLOWER, DEEPER BREATHS ENGAGES YOUR PARASYMPATHETIC SYSTEM AND HELPS YOU RELAX. BREATHING WITH YOUR CHEST, OR THORACIC BREATHING, IS MORE SHALLOW.
AHH... I SEE.

AREN'T THE MUSCLES USED FOR THORACIC BREATHING CALLED THE INTERCOSTAL MUSCLES?

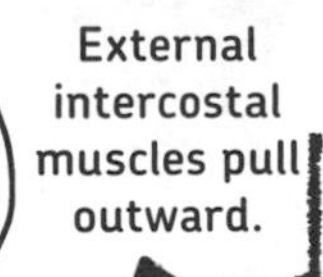
External intercostal muscles pull outward.

YES, THEY CAN HELP SUPPLEMENT BREATHING TO CREATE QUICK, SHALLOW BREATHS. THIS THORACIC BREATHING MIGHT OCCUR IN A STRESSFUL SITUATION, WHEN YOU NEED QUICK BURSTS OF OXYGEN.
THE CONTRACTION OF THE EXTERNAL INTERCOSTAL MUSCLES EXPANDS THE THORACIC CAVITY DURING INHALATION.
THEY RETURN TO THEIR ORIGINAL STATE IN EXHALATION.
FUN FACT! FOR PREGNANT WOMEN, BECAUSE THE GROWING UTERUS DECREASES THE MOBILITY OF THE DIAPHRAGM, THEIR RESPIRATORY RATE INCREASES TO MAINTAIN THE SAME LEVEL OF VENTILATION.

WELL, WHAT ABOUT *INTERNAL* INTERCOSTAL MUSCLES?
THEY DRAW *INWARD*...

THEY ARE USED WHEN BREATHING HEAVILY, TOO.
THEY CAN HELP PUSH OUT AIR DURING EXHALATION. YOU CAN FEEL YOURSELF USING YOUR EXTERNAL INTERCOSTAL MUSCLES WHEN INHALING...
AND YOUR INTERNAL INTERCOSTAL MUSCLES WHEN FORCEFULLY EXHALING.

NOW YOU'VE GOT ME DREAMING ABOUT *KALBI* SHORT RIBS! ♪

DID YOU CATCH ALL THAT?
IT IS ACTUALLY THE MUSCLES *BETWEEN* THE RIBS, BUT...

SWEET!
FOR NOW, THOUGH, HOW ABOUT SOME BREAKFAST?

CONTROLLING RESPIRATION

SO WE'VE TALKED ABOUT HOW YOU CAN CONSCIOUSLY CONTROL YOUR BREATHING.
BUT YOUR RESPIRATORY RATE CAN VARY EVEN WHEN YOU AREN'T THINKING ABOUT IT.
AH, JUST LIKE OTHER SYSTEMS IN THE BODY...

...THE RESPIRATORY SYSTEM AUTOMATICALLY ADAPTS TO RESPOND TO DIFFERENT SITUATIONS.

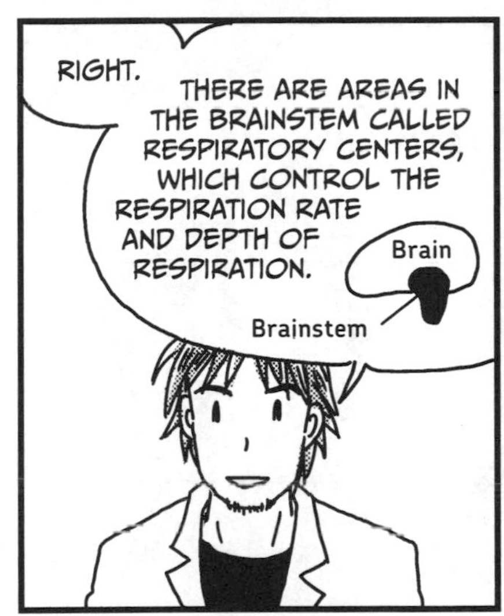
RIGHT.
THERE ARE AREAS IN THE BRAINSTEM CALLED RESPIRATORY CENTERS, WHICH CONTROL THE RESPIRATION RATE AND DEPTH OF RESPIRATION.
Brain
Brainstem

IF I REMEMBER CORRECTLY...
THE BRAINSTEM IS LIKE A KIND OF LIFE SUPPORT SYSTEM, ISN'T IT?
CARDIAC CYCLE
RESPIRATION
THAT'S RIGHT. OUR MOST BASIC FUNCTIONS ARE CONTROLLED THERE.

FOR EXAMPLE, INHALE AS MUCH AS YOU CAN... AND THEN HOLD YOUR BREATH. EVERYTHING YOU'VE DONE SO FAR WAS BY YOUR OWN INTENTION, RIGHT?
SUU
HAH HAH
Voluntary
PU-HAH!
BUT WHEN YOU REACH A LIMIT...
...YOU HAVE TO EXHALE, AND YOUR RESPIRATION RATE REALLY INCREASES.

I GET IT!
THAT'S BECAUSE THE CONTROL ROOM DECIDED THAT NOT BREATHING WAS A BAD IDEA AFTER ALL AND SOUNDED THE ALARM, RIGHT?
THAT'S RIGHT!

SO WHAT DO YOU SUPPOSE THE CONTROL ROOM SENSED WHEN IT GAVE THE INSTRUCTION TO EXHALE AND START BREATHING AGAIN?

I DIDN'T HAVE ENOUGH OXYGEN...

WELL, YOUR OXYGEN LEVELS WERE DECREASING, BUT IT WAS ACTUALLY THE BUILDUP OF CARBON DIOXIDE THAT CAUSED YOU TO TAKE ANOTHER BREATH.
Carbon Dioxide
Concentrated
Diluted
HAH HAH HAH
はっはっ
はー
すー
HAH SUU
CO2 CO2 CO2
CARBON DIOXIDE HAS INCREASED.
I SEE.

Carbon Dioxide and Respiratory Movement

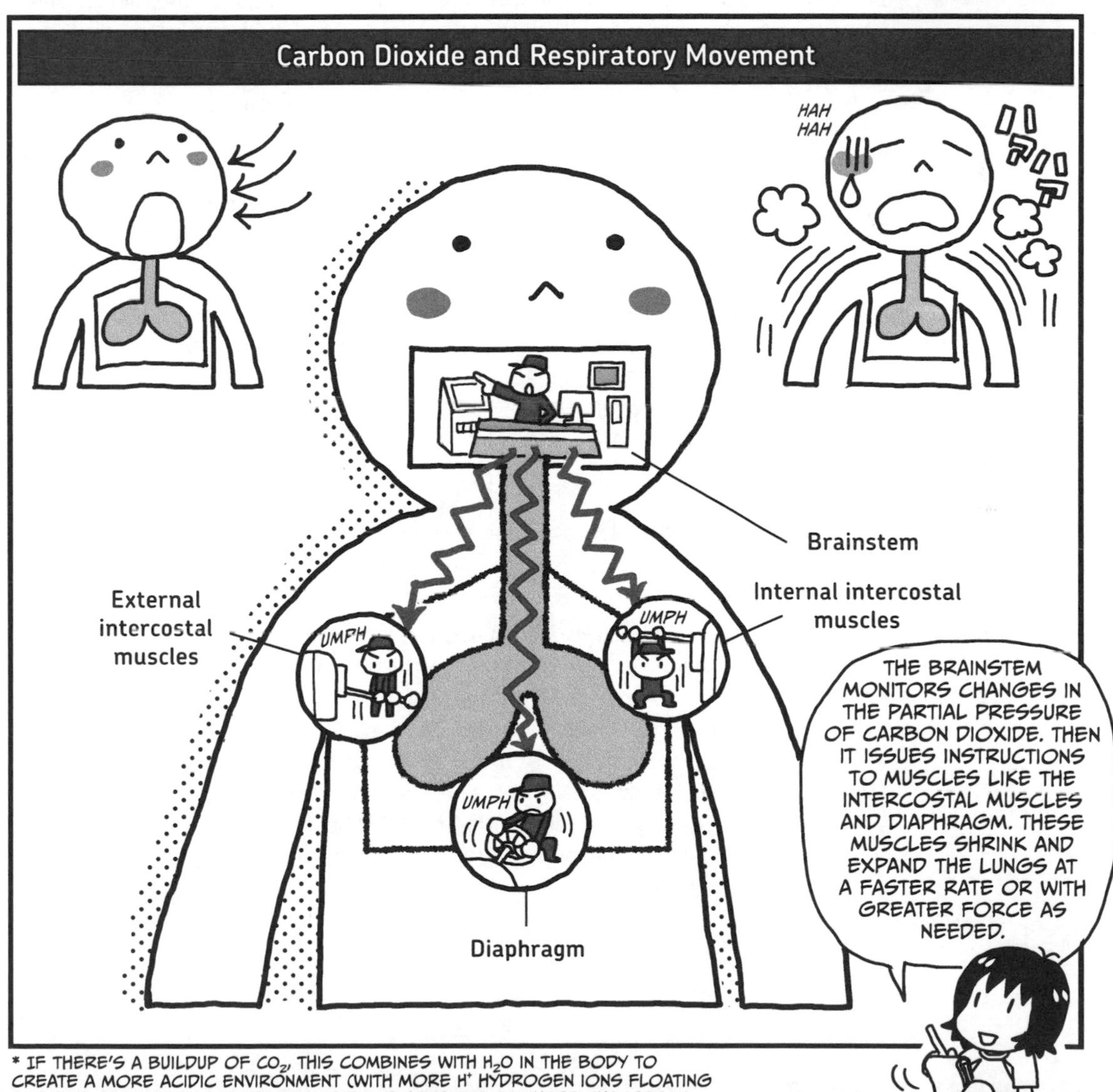

* IF THERE'S A BUILDUP OF CO_2, THIS COMBINES WITH H_2O IN THE BODY TO CREATE A MORE ACIDIC ENVIRONMENT (WITH MORE H^+ HYDROGEN IONS FLOATING AROUND). THE MEDULLA IN THE BRAINSTEM IS SENSITIVE TO THESE CHANGES IN CO_2 AND ACIDITY, AND IT CONTROLS THE RESPIRATORY RATE ACCORDINGLY.

EVEN MORE ABOUT THE RESPIRATORY SYSTEM!

The respiratory organs form a system that takes in oxygen to produce energy and disposes of the resulting carbon dioxide. Let's learn some more about the lungs, which are the main players in this system.

EXTERNAL AND INTERNAL RESPIRATION

So far, we've explained *ventilation*, which moves air into and out of the lungs. Next, we'll explain how the oxygen that's taken in by breathing is transported within the body and how carbon dioxide is disposed of at the same time.

This process is called *gas exchange*. Gas exchange occurs in two ways in the human body: through external respiration and internal respiration. *External respiration* happens in the lungs, where gases are exchanged between blood cells and alveoli (Figure 2-1). Alveoli are microscopic clusters of pockets located at the end of the airways (or bronchioles). This is how the blood receives oxygen from the air we breathe while expelling carbon dioxide.

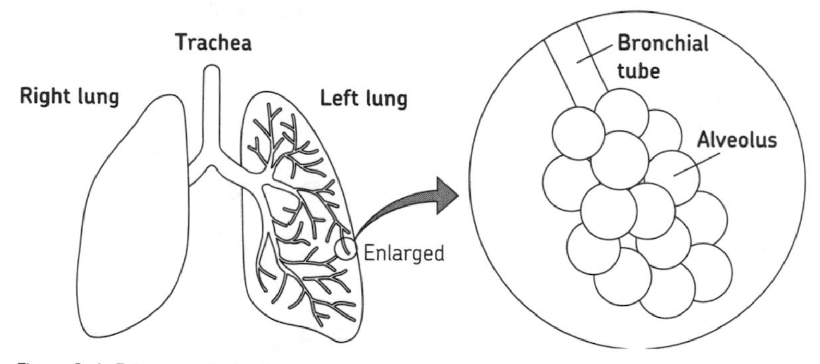

Figure 2-1: Enlargement of pulmonary alveoli

The gas exchange of external respiration is performed by each individual alveolus, right?

That's right. Although a single alveolus is tiny, there is an enormous number of them: approximately 700 million within a pair of human lungs. If all of the alveoli were spread out on a flat surface, they would cover an area approximately the size of a badminton court, or 100 square meters. The human body uses a lot of surface area for gas exchange!

Internal respiration is gas exchange that is performed within each cell as blood circulates through the tissues of the entire body. This is how oxygen is delivered through blood to cells in the body that need energy. You can refer to Figure 2-2 to see a full picture of how blood circulates throughout the body to perform internal and external respiration.

Figure 2-2: External and internal respiration

DID YOU KNOW?

Gas exchange in external and internal respiration is performed via a process called *diffusion* (see page 107). During diffusion, the gas spreads out from a concentrated region to a dilute region until ultimately, the concentration is even across both regions.

PARTIAL PRESSURES OF GASES IN THE BLOOD

Approximately how much oxygen and carbon dioxide are contained in the blood, and how does the proportion change during gas exchange? To answer this, we need to learn about the partial pressures of the gases.

When you have a mixture of gases, the partial pressure of a single gas is the pressure that gas would have if it took up all of the space occupied by the mixed gas. For example, the air around us contains a mixture of oxygen, nitrogen, carbon dioxide, and other gases. The partial pressure of oxygen is what you would get if you got rid of all the other gases besides oxygen and then measured the pressure of that oxygen within the same volume that the mixture previously filled. Partial pressure is represented by the letter *P* (for pressure), and the chemical name of the gas is added as a subscript. For example, the partial pressure of oxygen is P_{O_2}, and the partial pressure of carbon dioxide is P_{CO_2}.

Partial pressure is measured by millimeters of mercury, or mm Hg. If we take a volume of air under normal atmospheric pressure (1 atm, or 760 mm Hg), the partial pressure of oxygen within it is 160 mm Hg and the partial pressure of nitrogen is 600 mm Hg, as shown in Figure 2-3.

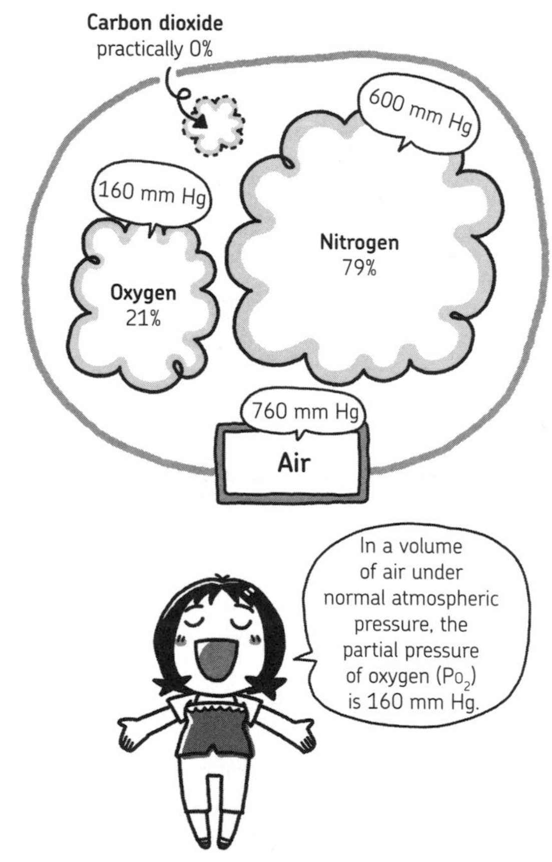

Figure 2-3: Components of the atmosphere and their partial pressures under 1 atmospheric pressure (760 mm Hg)

Now let's compare the partial pressures of gases in blood to the proportions of those gases in the air. Air is a mixed gas containing 21 percent oxygen, 0.03 percent carbon dioxide, and 79 percent other gases such as nitrogen, as shown in Figure 2-3. However, the human body does not use nitrogen at all, and the amount of carbon dioxide in the body is negligible. Therefore, the only thing we need to know here is the partial pressure of oxygen.

So what is the partial pressure of oxygen in the body? It depends on whether we're talking about oxygen in veins or arteries. The partial pressure of oxygen in arteries is denoted by P_{aO_2}, and the partial pressure of carbon dioxide in arteries is denoted by P_{aCO_2}. In veins, these are denoted by P_{vO_2} and P_{vCO_2}, respectively. In this notation, *a* stands for arteries and *v* stands for veins.

Now let's refer to Figure 2-4 and look at the transitions of the partial pressures in the body. The standard value for P_{aO_2} is 100 mm Hg. For P_{aCO_2}, it's 40 mm Hg, and for P_{vO_2}, it's 40 mm Hg.

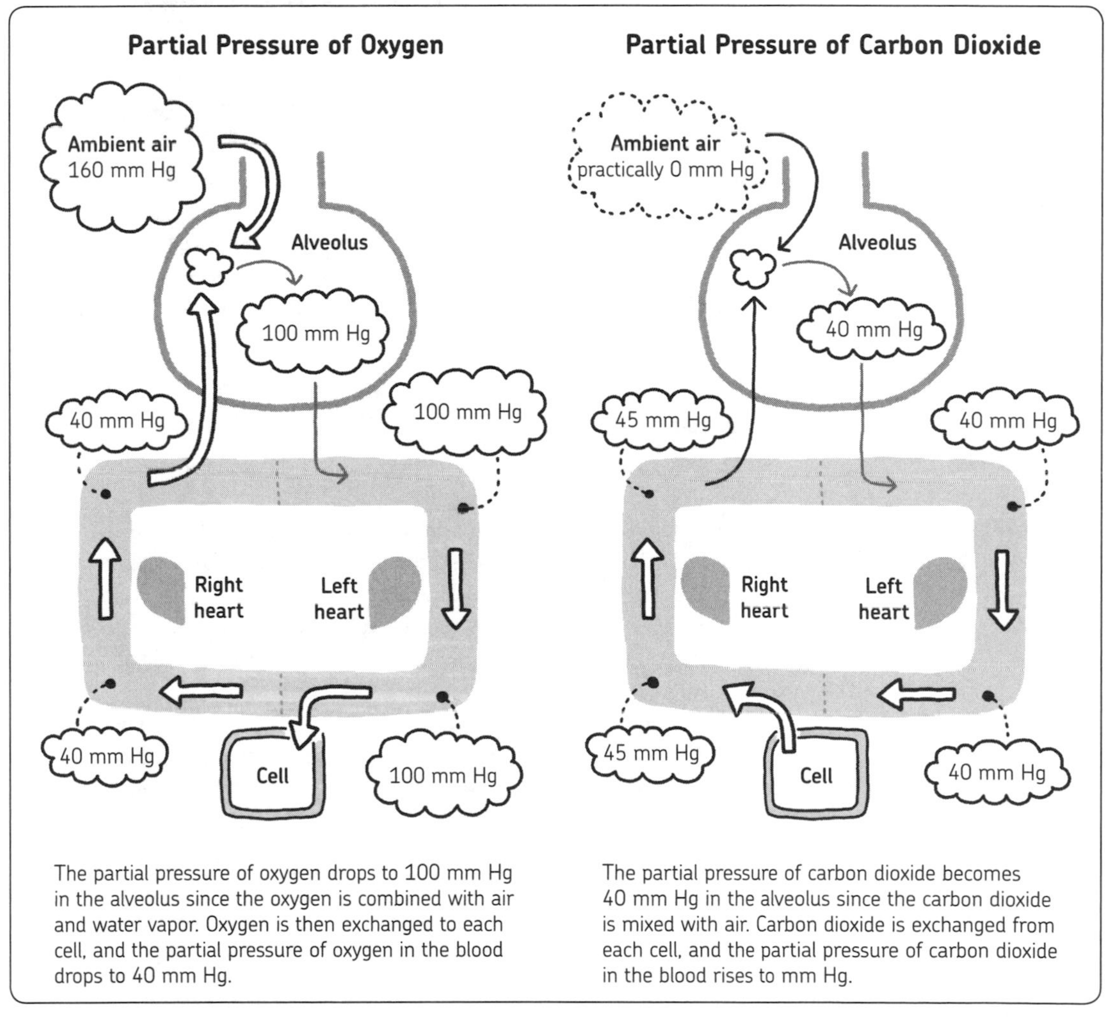

Figure 2-4: Changes in the partial pressures of oxygen and carbon dioxide in the body

DID YOU KNOW?

Carbon dioxide is readily exchanged from blood because there is practically no CO_2 in ambient air, so diffusion occurs easily and rapidly. The release of CO_2 (and therefore a decrease in P_{aCO_2}) is intimately related to pH in the body (see page 51).

ACIDOSIS AND ALKALOSIS

pH is a measure of a liquid's acidity or alkalinity. Like any other liquid, blood also has a pH, and its value changes due to respiration. Actually, the partial pressures of gases hold the key to understanding how respiration affects the body's pH. If the pH level of the blood exceeds a standard value, it will cause problems in the body. So how does the body regulate this value?

A pH of 7 is neutral. As the number decreases, a liquid becomes more acidic, and as the number increases, a liquid becomes more alkaline. The pH of the human body is approximately 7.4, which means the body is slightly alkaline. This pH is maintained at a nearly constant value. The mechanism that maintains a constant pH state within the body is called *homeostasis*.

If a problem occurs in certain bodily functions, the pH level may exceed the standard value range. The condition in which the pH level tends to be more acidic than the standard value is called acidosis, and the condition in which the pH level tends to be more strongly alkaline is called alkalosis.

Since a pH of 7 is neutral, is a body pH of 7.1 in a state of alkalosis?

No, no. Since acidosis and alkalosis are both relative to the standard value of pH 7.4, a pH of 7.1 is tending towards the acidic side. Therefore, that would be a case of acidosis (even though the pH level is still overall slightly alkaline).

Figure 2-5 shows acidosis and alkalosis relative to the body's pH. If bodily pH falls below 6.8 or rises above 7.8, there is a risk of death. However, since the body naturally is slightly alkaline, its pH rarely drops below 7 to become truly acdic.

So how do acidosis and alkalosis occur? The pH of the body can change based on the level of P_{aCO_2}—the two are closely related. When P_{aCO_2} is high, more acid is created in the body, and acidosis can occur. Conversely, when P_{aCO_2} is low, acid levels in the body decrease, and alkalosis can occur. Why does acid increase as P_{aCO_2} increases? This is because the dissolution of carbon dioxide in water produces H^+ ions, which make things more acidic.

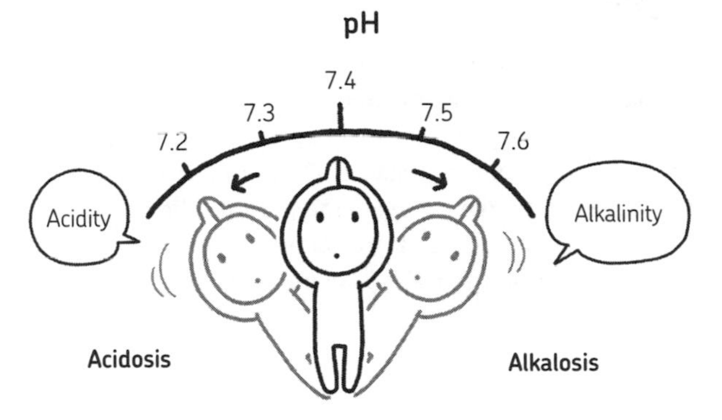

Figure 2-5: Acidosis occurs when the body's pH tends to be acidic, and alkalosis when its pH tends to be alkaline.

Carbon dioxide dissolves in water . . . ?

Well, think of a carbonated beverage. The carbonation of a carbonated beverage is just carbon dioxide dissolved in water.

DID YOU KNOW?

This is the chemical equation to describe how carbon dioxide dissolves in water within the body:

$$H_2O + CO_2 \text{ « } H^+ + HCO_3^-.$$

If the concentration of this hydrogen ion (H^+) increases in an aqueous solution (such as blood), the pH will tend toward the acidic side.

Incomplete respiration (or *hypoventilation*) results in too much carbon dioxide in the body. More carbon dioxide creates a more acidic environment, which can in turn cause acidosis.

Hyperventilation is a condition in which ventilation is excessive. Since it causes a state in which a lot of carbon dioxide is being expelled, the P_{aCO_2} level will decrease, thereby causing the pH of the body to become more alkaline. Acidosis and alkalosis can also be caused by metabolic abnormalities (see "ATP and the Citric Acid Cycle" on page 74).

HOW THE LUNGS WORK

Now let's take a look at the lungs. Pulmonary function testing measures the amount of air you can inhale and exhale and the amount of force you need to exert to do this. The results are represented in a graph called a *spirogram* (like the one in Figure 2-6) that shows the volume of air at different stages of inspiration or expiration. The initial small periodic curve indicates the interval when the person is breathing normally. The peaks are locations when inspiration ends (resting inspiratory volume), and the troughs are locations when expiration ends (resting expiratory volume). The difference between these levels is the resting tidal volume.

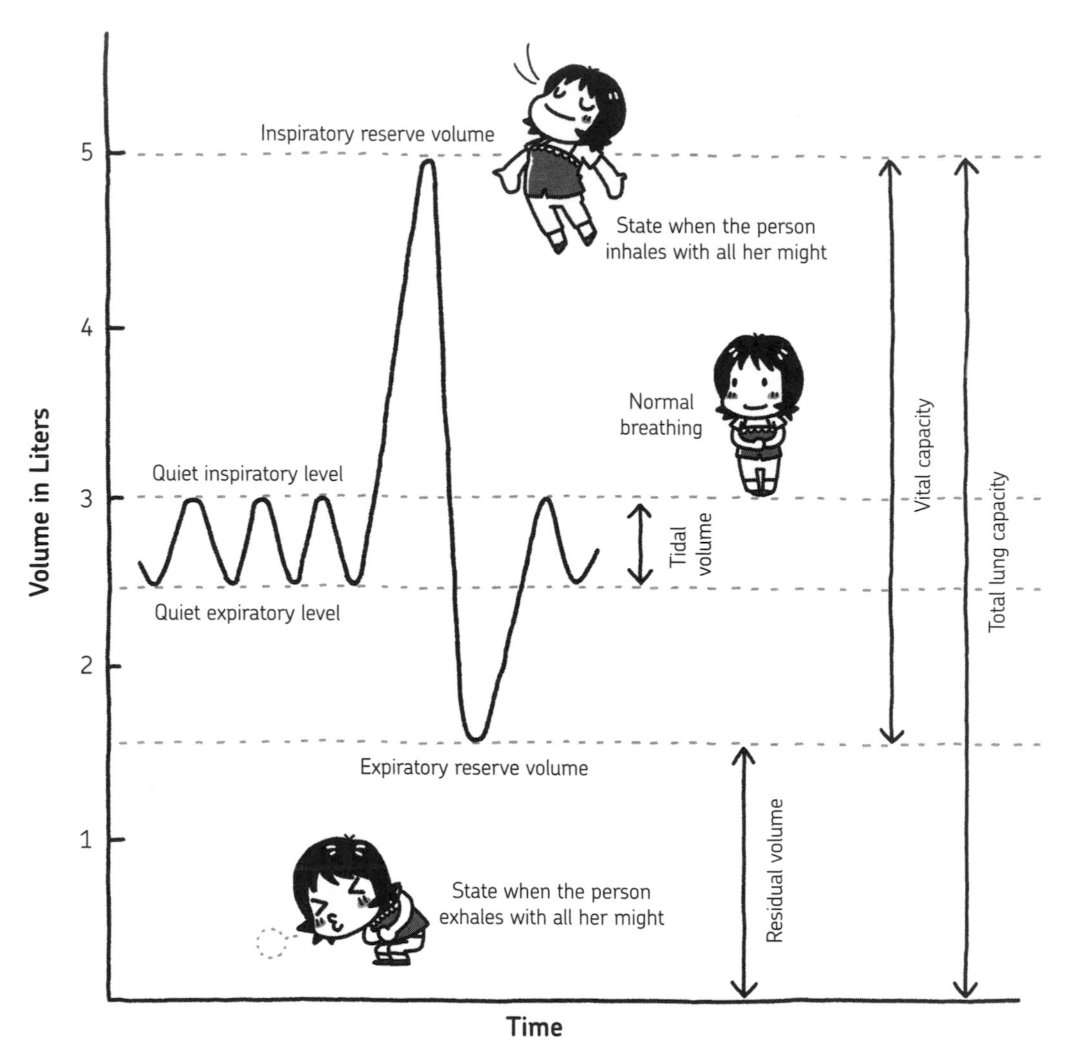

Figure 2-6: Understanding the output of a spirogram

Following this, you see a marked peak and trough. The highest peak (where the person inhaled with all her might) is the inspiratory reserve volume, and the deepest trough (where the person exhaled with all her might) is the expiratory reserve volume. As shown in Figure 2-6, the *vital capacity* is the difference in volume between maximum inhalation and complete exhalation.

We must not forget *residual volume* here. Even if a person exhales with all her might, since the lungs do not become completely flat and the trachea and bronchial tubes also do not flatten out, a fixed volume of air remains in those locations. This volume is called the residual volume.

The sum of the residual volume and vital capacity, which is the entire capacity of the lungs, is called the total lung capacity:

$$\text{Total Lung Capacity} = \text{Vital Capacity} + \text{Residual Volume}$$

My vital capacity is 3500 milliliters. That's a lot, isn't it?

You're right. The standard for a woman is 2000 to 3000 milliliters. Your number reflects your training as a marathon runner. The vital capacity for a man is approximately 3000 to 4000 milliliters. Vital capacity tends to be greater for people who have a larger physique.

3 THE DIGESTIVE SYSTEM

DIGESTION, METABOLISM, AND THE MULTI-TALENTED LIVER

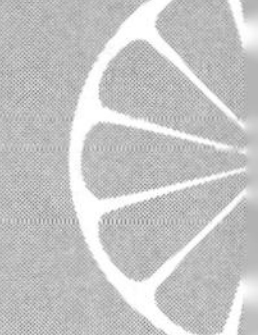

THE ALIMENTARY CANAL

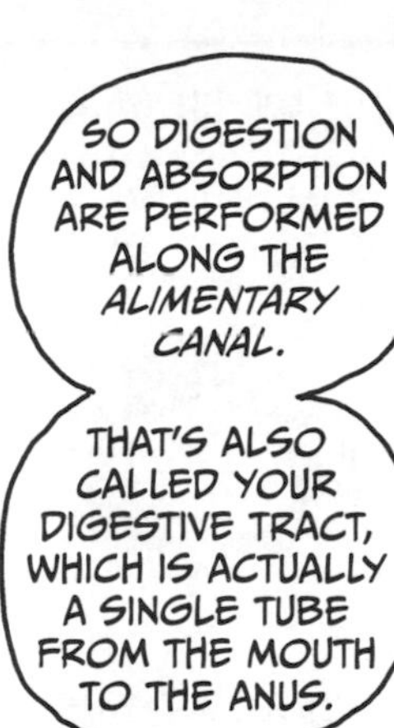
SO DIGESTION AND ABSORPTION ARE PERFORMED ALONG THE *ALIMENTARY CANAL*.
THAT'S ALSO CALLED YOUR DIGESTIVE TRACT, WHICH IS ACTUALLY A SINGLE TUBE FROM THE MOUTH TO THE ANUS.

DON'T SAY "ANUS" WHEN I'M EATING!

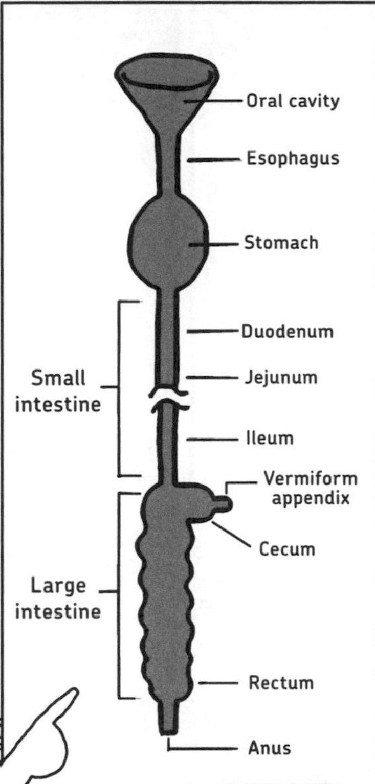
Oral cavity
Esophagus
Stomach
Duodenum
Jejunum
Small intestine
Ileum
Vermiform appendix
Cecum
Large intestine
Rectum
Anus

FIRST, *MASTICATION*—OR CHEWING—IS PERFORMED IN THE MOUTH, RIGHT?

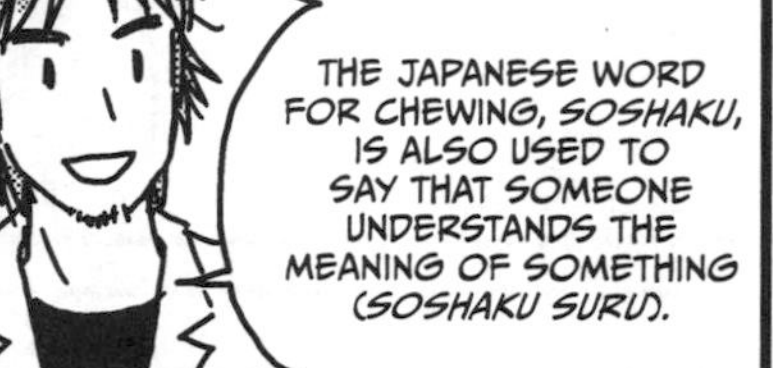
THAT'S RIGHT. IT GRINDS THINGS UP REALLY WELL.
THE JAPANESE WORD FOR CHEWING, *SOSHAKU*, IS ALSO USED TO SAY THAT SOMEONE UNDERSTANDS THE MEANING OF SOMETHING (*SOSHAKU SURU*).

THERE'S ALSO A JAPANESE EXPRESSION *NOMIKOMI GA HAYAI* (LITERALLY "QUICK SWALLOWING") THAT MEANS "QUICK TO LEARN."

I TAKE IT YOU'VE HEARD THAT ONE OFTEN, THEN?
THE MOUTH IS THE STARTING POINT OF THE DIGESTIVE TRACT, RIGHT? SO DO YOU KNOW WHAT WORK IT DOES?

THE TEETH, JAW, AND TONGUE ARE USED FOR CHEWING...

AH! AND SALIVA IS ALSO PRODUCED IN THE MOUTH.
RIGHT. FOOD IS BLENDED WITH SALIVA TO MAKE IT EASY TO SWALLOW...
AND SALIVA HAS DIGESTIVE ENZYMES THAT CONVERT NUTRIENTS IN FOOD INTO A FORM THAT THE BODY CAN ABSORB.
UGH.
THERE ARE SO MANY DIGESTIVE ENZYMES...
PTYALIN
TRYPSIN
SUCRASE
PANCREATIC AMYLASE
MALTASE
PANCREATIC LIPASE
PEPSIN
CHYMOTRYPSIN
LIPASE

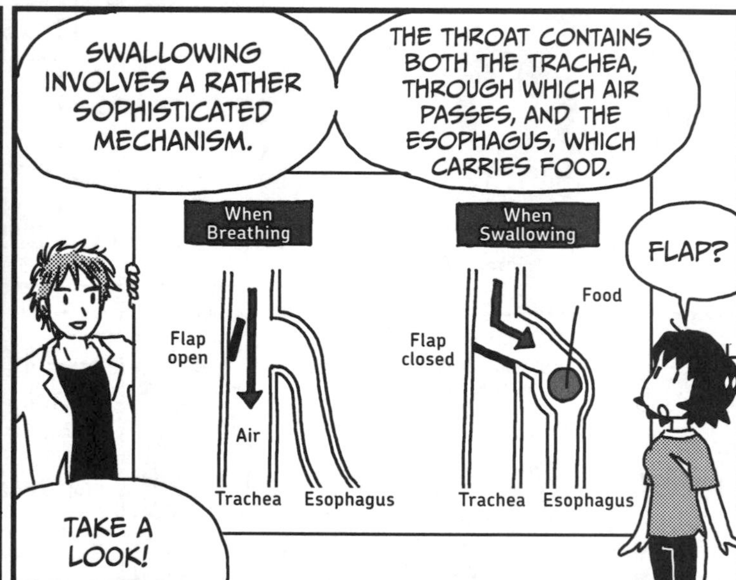

* ASPIRATING FOOD CAN CAUSE LUNG INJURY BECAUSE FOOD CARRIES THE GASTRIC ACID AND BACTERIA FOUND IN THE DIGESTIVE TRACT.

THE ESOPHAGUS AND THE STOMACH

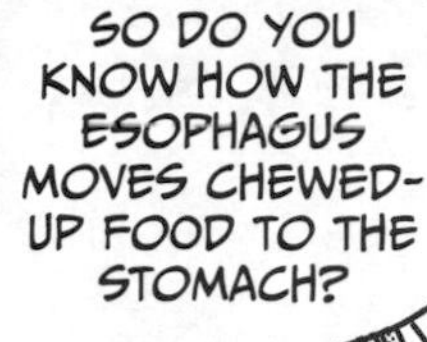

* PEPSINOGEN IS A *ZYMOGEN*, OR A *PROENZYME* (AN INACTIVE PRECURSOR TO AN ENZYME). PEPSINOGEN IS ACTIVATED AND TRANSFORMED INTO PEPSIN BY HYDROCHLORIC ACID.

THE DUODENUM AND THE PANCREAS

THE SMALL AND LARGE INTESTINES
WOW, DIGESTION SEEMS LIKE AN AWFUL LOT OF WORK.
NEXT, THOSE NUTRIENTS ARE FINALLY ABSORBED, AREN'T THEY?
OH!
INDEED!
DO YOU KNOW WHICH ORGAN IS RESPONSIBLE FOR THAT WORK?
THE SMALL INTESTINE!
MORE SPECIFICALLY, IT'S THE DUODENUM, JEJUNUM, AND ILEUM!
TOGETHER, THESE MAKE UP THE SMALL INTESTINE...
AND ABSORPTION OF DIGESTED FOOD OCCURS IN THE CELLULAR SURFACE OF THE MUCUS MEMBRANE OF THE SMALL INTESTINE.
Stomach
Duodenum
Jejunum
Ileum
Large intestine
Intestinal wall
PERISTALTIC MOTION, WHICH WE SKIPPED EARLIER, ALSO OCCURS HERE.
SO THE INTESTINE PERFORMS PERISTALTIC MOTION LIKE THE ESOPHAGUS DOES, RIGHT?
YES, PERISTALSIS IS AN EARTHWORM-LIKE MOVEMENT.
IT MOVES THINGS ALONG TOWARD THE ANUS.
Constrictions behind the lump cause it to move forward.

PERSONALLY, I THINK OF TOOTHPASTE BEING SQUEEZED OUT OF A TUBE.

THAT'S ALSO PRETTY CLOSE.

BY THE WAY, BECAUSE OF PERISTALTIC MOTION, YOU CAN EAT A MEAL EVEN IF YOU'RE STANDING ON YOUR HEAD OR IN A WEIGHTLESS STATE. AND OF COURSE, YOU CAN ALSO POOP IN SPACE, TOO.

WOW. WE'VE COVERED ALL OF THE DIGESTIVE ORGANS.

YOU FORGOT POOPING!
LET'S SAY DEFECATION, OKAY?!

MOISTURE IS ABSORBED BY THE LARGE INTESTINE.
SO POOP WITH A PROPER CONSISTENCY CAN BE FORMED.

IF YOUR LARGE INTESTINE DOESN'T ABSORB MOISTURE PROPERLY, YOUR POOP WILL BE RUNNY.
SHE'S NO BETTER THAN HE IS...

ENOUGH ALREADY. LET'S SAY SOFT STOOL!
OK OK...
COUGH COUGH
WE'RE ALMOST AT THE END OF THE LINE: THE ANUS.

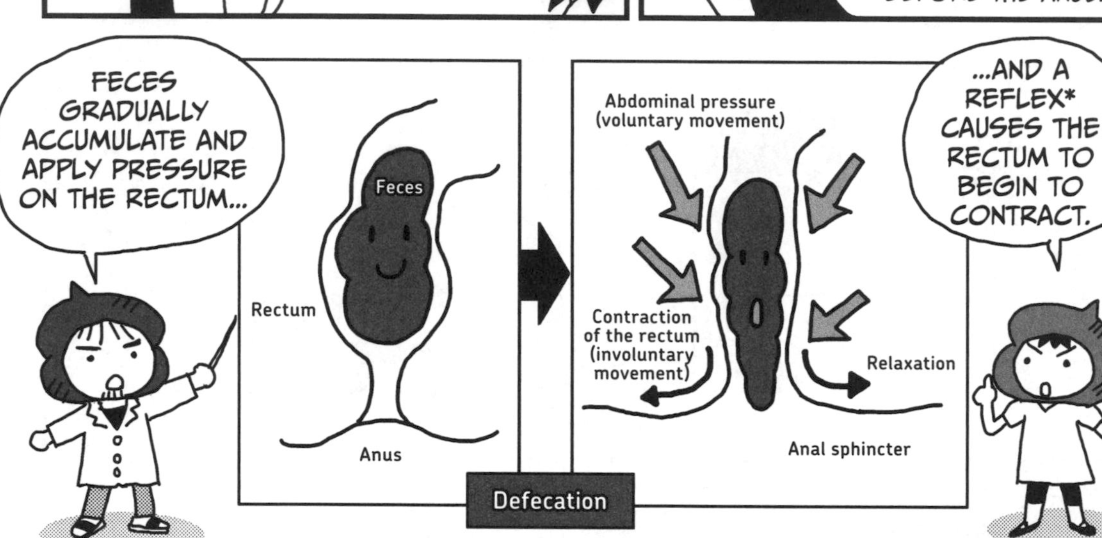

* A REFLEX IS AN INVOLUNTARY MOVEMENT IN RESPONSE TO A STIMULUS.

THE THREE MAJOR NUTRIENTS

FIRST, LET'S START WITH CARBOHYDRATES.
CARBOHYDRATES INCLUDE GLUCOSE, FRUCTOSE, GALACTOSE, LACTOSE, MALTOSE, SUCROSE, AND STARCH.
THE ONLY ENERGY SOURCE FOR THE BRAIN IS GLUCOSE!
THAT'S WHY SWEETS ARE INDISPENSABLE DURING STUDY BREAKS!
THE FOUNDATION OF CARBOHYDRATES IS GLUCOSE!
EVEN IF YOU FAST FOR A COUPLE OF DAYS, IT'S UNLIKELY THAT YOUR BRAIN'S ENERGY SOURCE WILL BE COMPLETELY DEPLETED.
BUT WHEN YOU'RE REALLY HUNGRY, YOUR MIND BEGINS TO WANDER.

YOU BECOME A BIT OF A BLOCKHEAD.
HARD AS A SUGAR CUBE!
BY THE WAY, TABLE SUGAR IS SUCROSE.

LET'S LOOK AT ALL THE DIFFERENT KINDS OF CARBOHYDRATES.

Types of Carbohydrates

Monosaccharides*

Glucose   Fructose   Galactose

Disaccharides**

Sucrose	Lactose	Maltose
Glucose + fructose	Glucose + galactose	Glucose × 2

Polysaccharides***

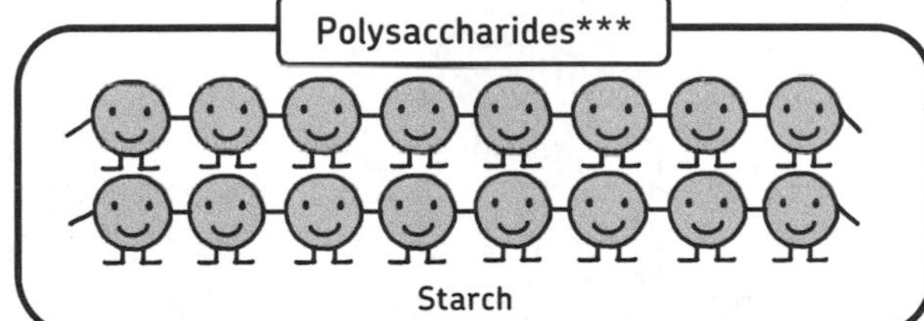

Starch

* MONOSACCHARIDES: MOST EASILY ABSORBED BY THE BODY
** DISACCHARIDES: A FORM IN WHICH TWO MONOSACCHARIDES ARE JOINED TOGETHER
*** POLYSACCHARIDES: A FORM IN WHICH MANY TYPES OF MONOSACCHARIDES ARE JOINED TOGETHER

** ESSENTIAL FATTY ACIDS ARE UNSATURATED FATTY ACIDS THAT THE BODY CANNOT PRODUCE BY ITSELF, SO MUST ABSORB FROM FOOD.

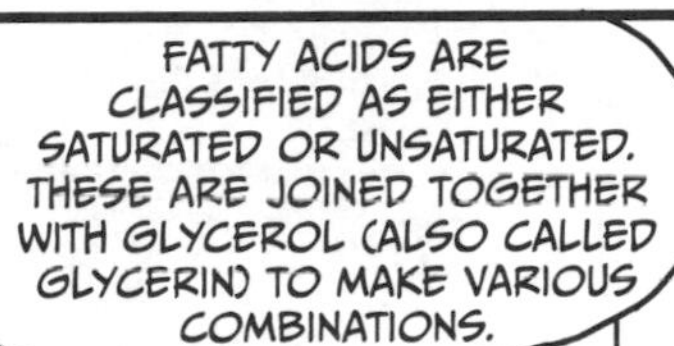

* NEUTRAL FATS ARE CLASSIFIED BY FATTY ACIDS THAT FORM THEM, AS WELL AS THE NUMBER OF CARBON ATOMS AND THE CONNECTIONS BETWEEN THEM. EVEN BEYOND TRIGLYCERIDES, THERE CAN ALSO BE MONO- AND DIGLYCERIDES.

LAST, WE'RE GOING TO COVER PROTEINS, RIGHT?
PROTEINS ARE IMPORTANT NUTRIENTS FOR THE STRUCTURAL COMPONENTS OF THE BODY, SUCH AS MUSCLE, BONE, SKIN, AND BLOOD.
RIGHT.
OF COURSE, THEY CAN ALSO BE USED FOR ENERGY.
バーーン
BAM!
BAM!
Pepsin
バキッ
Pepsin
PEPSIN'S BACK! AND HE'S ON A RAMPAGE!
Protein
Amino acid
Peptide
PROTEINS CONSIST OF MANY AMINO ACIDS LINKED TOGETHER.
SMALL NUMBERS OF LINKED AMINO ACIDS ARE OFTEN CALLED PEPTIDES.
PEPSIN BREAKS DOWN PROTEINS INTO THOSE SMALLER PIECES!
RIGHT! DIGESTION CAN PROCEED!
?
HUH?
PROTEINS AND AMINO ACIDS SEEM A LITTLE LIKE TOY BUILDING BLOCKS.
Pepsin

FIRST, LET'S IMAGINE THIS BUILDING MADE OF BLOCKS IS A PROTEIN.
MANY DIFFERENT BLOCKS ARE JOINED TOGETHER.
THE FIRST PART OF THE PROCESS OF METABOLISM (CALLED *CATABOLISM*) CONSISTS OF BREAKING THIS BUILDING INTO PIECES.
WHEN THIS OCCURS DURING DIGESTION, WE END UP WITH INDIVIDUAL PARTS.
SO EACH INDIVIDUAL BLOCK IS AN AMINO ACID?
Amino acid
THIS BLOCK IS DIFFERENT.
IT'S GREAT THAT YOU NOTICED THAT.
THAT IS AN *ESSENTIAL* AMINO ACID.
LIKE THE ESSENTIAL FATTY ACIDS WE TALKED ABOUT EARLIER...
WE CAN'T CREATE THEM OURSELVES, SO WE ABSORB THEM FROM FOOD, RIGHT?
THAT'S RIGHT!
THE SECOND PHASE OF METABOLISM IS CALLED *ANABOLISM*. IT'S THE REASSEMBLY OF THE INDIVIDUAL BLOCKS (AMINO ACIDS) INTO MORE COMPLEX PROTEINS.
WHEN ANABOLISM OCCURS, NEW STRUCTURES ARE CREATED THAT ARE DIFFERENT FROM THE ORIGINAL CONFIGURATION (LIKE MAKING AN AIRPLANE OR ROBOT FROM A PILE OF BLOCKS)!
I GET IT...
MORE ACCURATELY, THOSE PARTS MIGHT BE USED TO CREATE MUSCLE, BONE, OR SKIN, FOR EXAMPLE!

EVEN MORE ABOUT THE DIGESTIVE SYSTEM!

Let's look at all of the digestive organs again! These include the organs of the alimentary canal (or digestive tract) from the mouth to the anus—as well as the liver, gallbladder, and pancreas. If we liken the sequence of processes included in digestion and absorption to a factory, it would look a little like this.

THE DIGESTIVE SYSTEM IN ACTION

❸ **Deglutition (swallowing)**
The swallowing of food that was ground down into finer pieces in the mouth is called *deglutition*. Food passes through the esophagus and enters the stomach.

❶ **Mastication (chewing)**
Mastication is collaborative work performed by the teeth, jaw, and tongue.

❷ **Saliva**
Saliva moistens the food. It contains digestive enzymes.

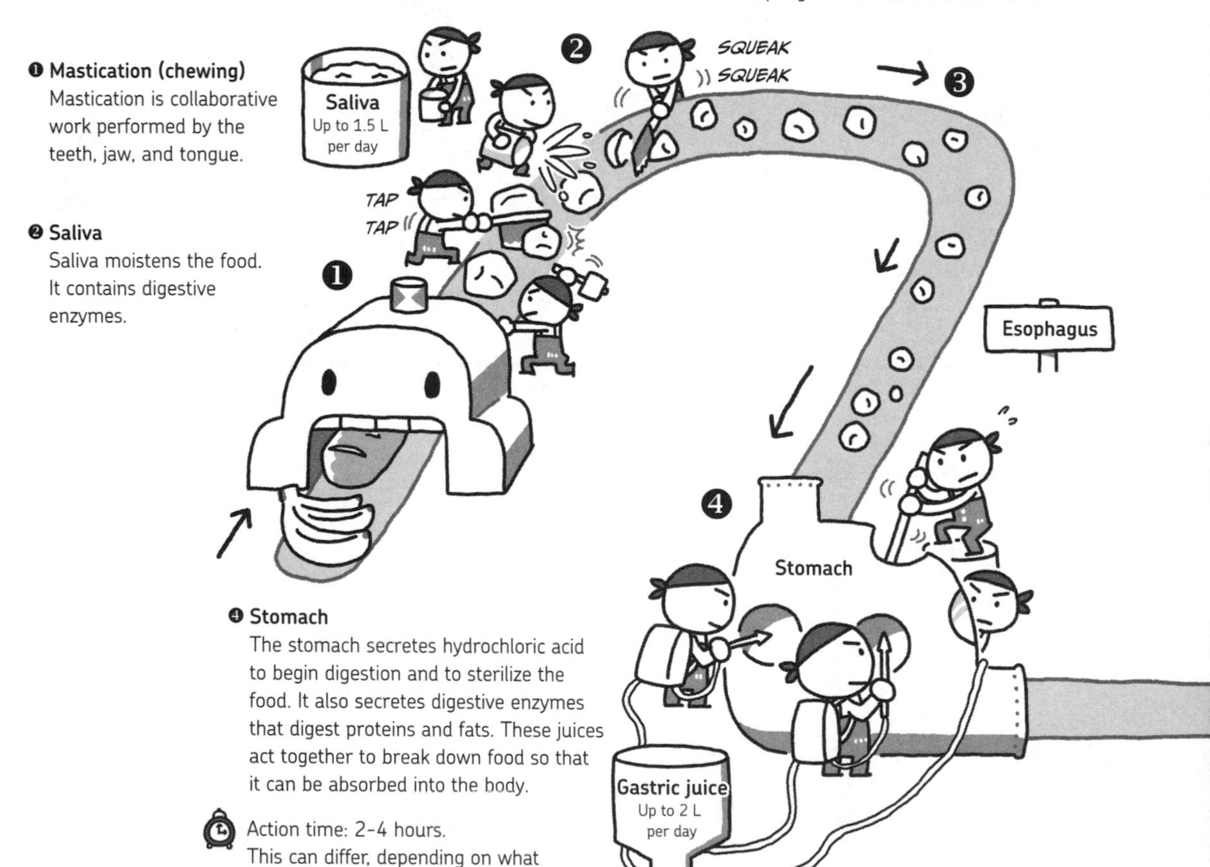

❹ **Stomach**
The stomach secretes hydrochloric acid to begin digestion and to sterilize the food. It also secretes digestive enzymes that digest proteins and fats. These juices act together to break down food so that it can be absorbed into the body.

Action time: 2–4 hours.
This can differ, depending on what you've eaten. Foods that have a lot of protein take longer to digest.

❺ Duodenum
Alkaline digestive fluids are mixed into the chyme from your stomach to neutralize the stomach acids before they enter the intestines. Pancreatic juice secreted from the pancreas also contains digestive enzymes to break down proteins and fats. Bile secreted from the gallbladder helps digest lipids (see Figure 3-7 on page 80).

❻ Small Intestine
Digestive fluids that carry out the final stage of digestion are secreted, and nutrients are steadily absorbed at the same time from nutrient absorption cells that line the walls of the small intestine. The length of the small intestine is approximately 6 to 8 meters in an adult.

 Action time: Approximately 3 to 5 hours

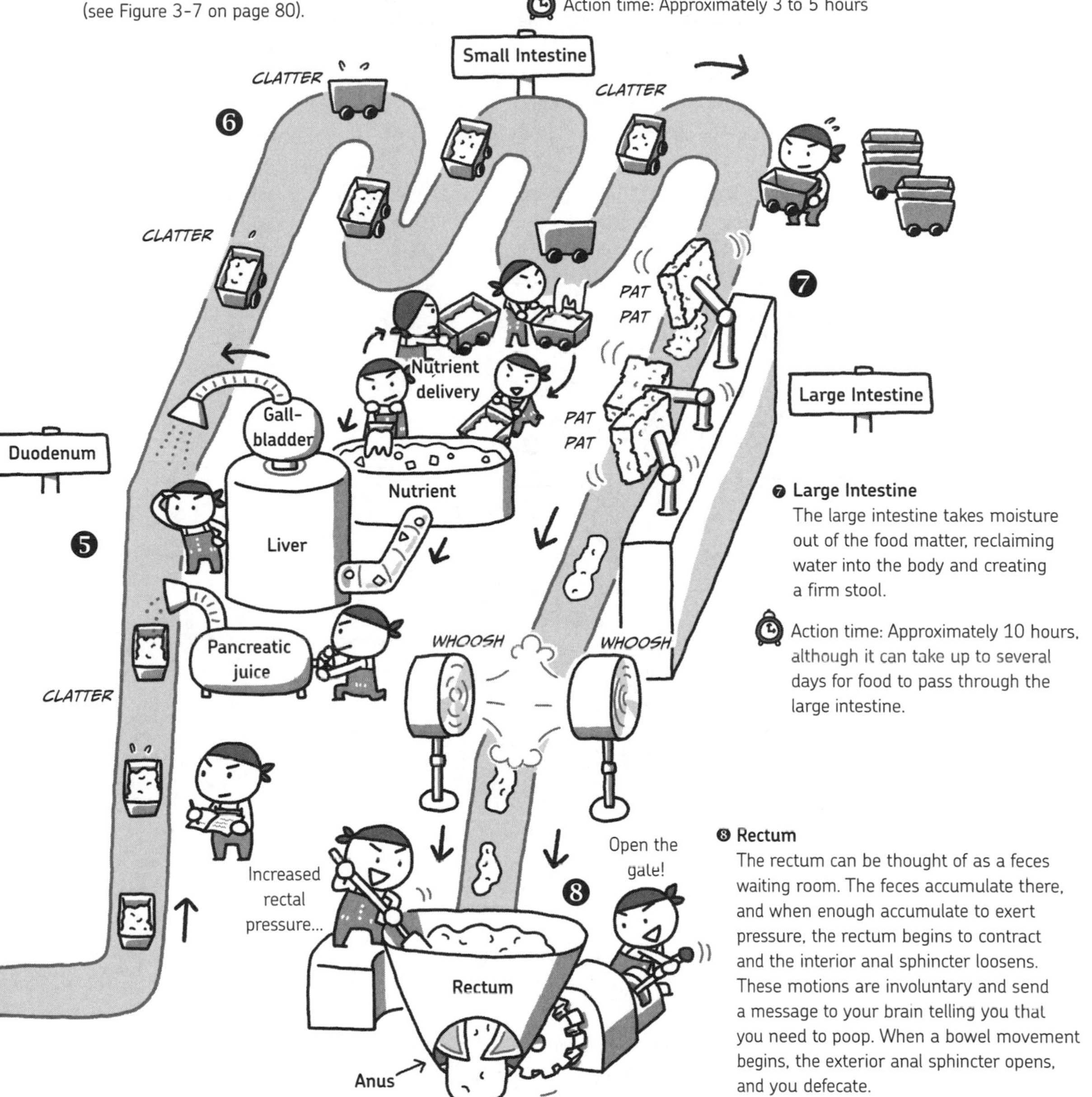

❼ Large Intestine
The large intestine takes moisture out of the food matter, reclaiming water into the body and creating a firm stool.

Action time: Approximately 10 hours, although it can take up to several days for food to pass through the large intestine.

❽ Rectum
The rectum can be thought of as a feces waiting room. The feces accumulate there, and when enough accumulate to exert pressure, the rectum begins to contract and the interior anal sphincter loosens. These motions are involuntary and send a message to your brain telling you that you need to poop. When a bowel movement begins, the exterior anal sphincter opens, and you defecate.

ATP AND THE CITRIC ACID CYCLE

Our bodies get energy from the foods and nutrients we take in. Our bodies synthesize or decompose such food and nutrients, and the reactions that carry out these processes are called *metabolism*. In this section, we will explain metabolism in a little more detail.

First, let's look at the process that produces energy by burning nutrients that are absorbed. Although we say "burning," the energy source is not being set on fire inside the body. Instead, the energy is produced by a chemical reaction called *oxidation*.

Oxidation extracts energy from nutrients (carbohydrates, fats, and proteins) by using oxygen. These energy sources are oxidized to form *ATP (adenosine triphosphate)*. We walk, digest, and carry out other activities using the energy that is produced by decomposing this ATP.

ATP decomposition is the energy source for all the activity that goes on inside cells, so it is performed inside every cell in the human body. The energy obtained from ATP decomposition is ultimately released as heat, as shown in Figure 3-1.

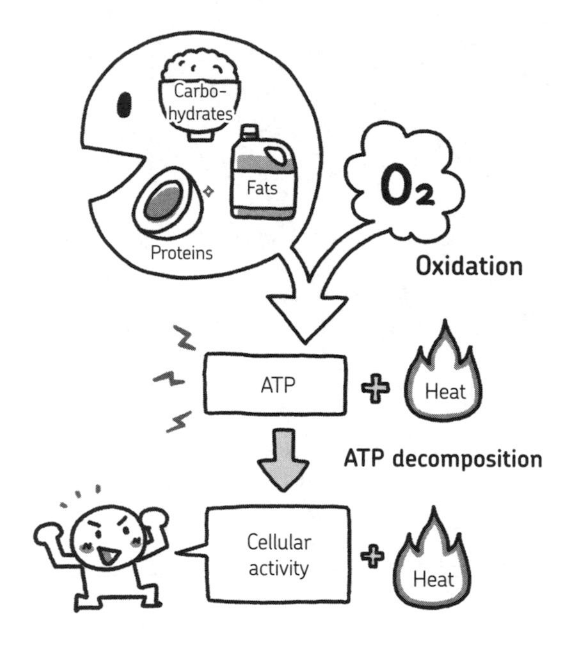

Figure 3-1: ATP decomposition

Where does ATP decomposition happen in the body?

ATP decomposition is the energy source for all the activity that goes on inside cells, so it is performed inside every cell in the human body. The energy obtained from ATP decomposition is ultimately released as heat, as shown in Figure 3-1.

The series of chemical reactions that burn nutrients to create ATP is called the *citric acid cycle* (see Figure 3-2). You don't have to remember the specific reactions in the citric acid cycle for each nutrient. For now, just remember what the entire citric acid cycle accomplishes.

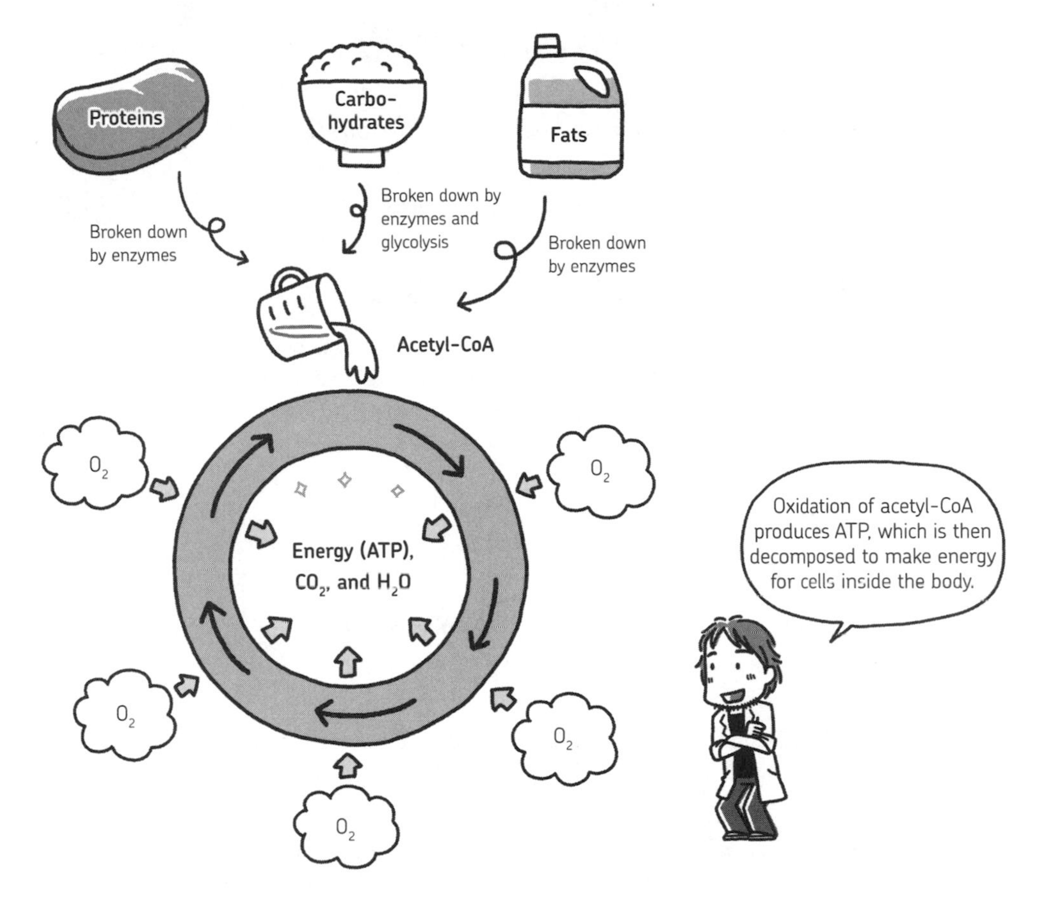

Figure 3-2: The citric acid cycle

So how does the flow of energy production occur in the citric acid cycle? First, each nutrient is broken down by enzymes. Proteins and fats are broken down into amino acids and fatty acids, which are eventually transformed into a molecule called acetyl-CoA, which then feeds into the citric acid cycle. For carbohydrates, the process is a little more complex. Carbohydrates are broken down into glucose, which is then broken down further through *glycolysis*. Glycolysis produces a small amount of energy (2 ATP per glucose molecule) and yields pyruvate, which is eventually changed into acetyl-CoA.

Once the three major nutrients are broken down into acetyl-CoA, the acetyl-CoA enters the citric acid cycle. Enzymes use oxygen to extract energy by causing a successive series of oxidation reactions. This produces high-energy molecules that are ultimately transformed into ATP through another series of reactions called the *electron transport chain*. Carbon dioxide and water are also by-products of the citric acid cycle and the electron transport chain.

This cycle is an integral part of aerobic metabolism. Eventually, the citric acid cycle and other related pathways create about 20 times more ATP than does glycolysis alone.

DID YOU KNOW?

The citric acid cycle is also called the TCA cycle or the Krebs cycle, named after Hans Adolf Krebs, who received a Nobel Prize in medicine for his work on the subject.

DIGESTIVE FLUIDS AND DIGESTIVE ENZYMES

We talked a bit about digestive fluids and enzymes earlier. Let's go over them in more detail now. If you check the overall picture of the digestive organs shown on page 72, it will be easier to visualize what is going on.

Different organs secrete different digestive fluids, right?

Indeed. Can you name them in order? Start with the mouth.

Well, there's saliva in the mouth, gastric juice in the stomach, pancreatic juice and bile in the duodenum, and intestinal fluids in the small intestine.

That's right. In one day you secrete around 8 liters of digestive fluids! You'd think you'd get dehydrated disgorging all that liquid, huh? Well, don't worry. The moisture that is contained in the digestive fluids is absorbed by the alimentary canal as soon as the digestive fluids are secreted. As a result, you don't get dehydrated.

Most digestive fluids contain digestive enzymes—bile is the only one that does not. But it still can be called a digestive fluid since it aids digestion, acting like soap to disperse and emulsify fats. Bile is the bitter yellow liquid that comes up when you vomit violently. If there's nothing else to throw up, bile will be discharged. Bile is formed in part from the breakdown of old red blood cells (see "The Liver's Role in Metabolism and Digestion" on page 78).

Next, let's cover digestive enzymes. Your body cannot easily absorb food in its original form. Digestive enzymes play a major role in changing that food to a state that can be absorbed as nutrients by the body.

Is that why I have to memorize them?

Don't worry. I'm going to narrow them down to just three main types of digestive enzymes. There are rules that make it easier to remember the names.

It is customary for enzyme names to end in *-ase*. The Latin name for starch is *amylum*, so the enzyme that breaks down carbohydrates is called *amylase*. *Protease* is the enzyme that breaks down protein, and *lipase* is the enzyme that breaks down fats, or lipids (see Figure 3-3).

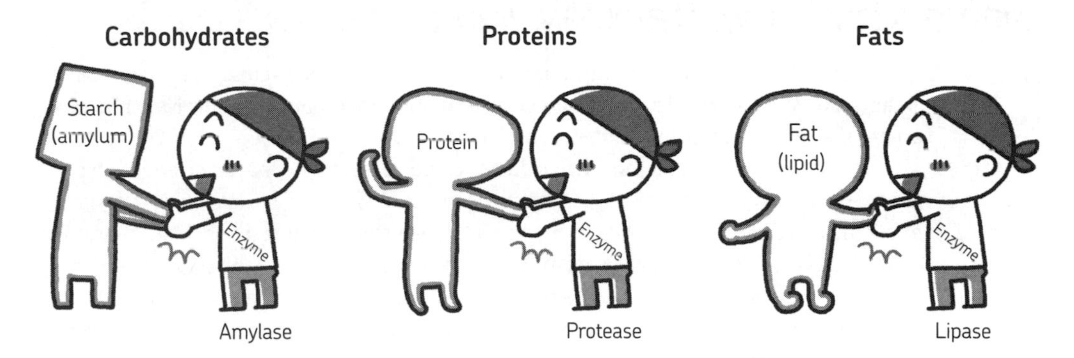

Figure 3-3: Three main types of digestive enzymes

Digestive enzymes can be broadly classified into three types in this way. Amylase and protease can also be classified further into several types. Since there are only a few important digestive enzymes, you'll be able to remember them if we organize them properly, as in Table 3-1.

TABLE 3-1: MAIN DIGESTIVE ENZYMES AND THEIR EFFECTS

	Enzymes for breaking down carbohydrates	**Enzymes for breaking down proteins**	**Enzymes for breaking down fats**
Saliva	*Salivary amylase* starch → maltose		
Gastric juice		*Pepsin* proteins → peptides*	
Pancreatic juice	*Pancreatic amylase, etc.* starch → maltose	*Trypsin, chymotrypsin* proteins → peptides or amino acids	*Pancreatic lipase* fats (lipids) → fatty acids + glycerin
Intestines	*Sucrase, etc.* sucrose, lactose, etc. → monosaccharide	*Erepsin* proteins or peptides → amino acids	

* Peptides are short, linked-together chains of amino acids, which have fewer molecules than proteins.

There sure are a lot, and some enzymes don't end in *-ase*.

If only it were that easy! Just remember that the enzymes without *-ase* break down proteins.

THE LIVER'S ROLE IN METABOLISM AND DIGESTION

Finally, let's talk about the liver, which performs important work involved in digestion and metabolism. You probably know that the liver is an internal organ that's involved in a wide variety of activities—up to 500 different functions!

The liver receives blood from two kinds of blood vessels: the *hepatic arteries*, which carry arterial blood containing oxygen from the aorta, and the *portal vein*, which carries venous blood containing nutrients that were absorbed from the intestines. More than two-thirds of the liver's blood supply is from the portal vein, supplying it with about half of the oxygen the liver needs, as well as the nutrients critical for various metabolic activities.

Since the liver is involved in so many different activities, I'll explain just a few of the major activities one at a time. The function of the liver is shown in Figure 3-4.

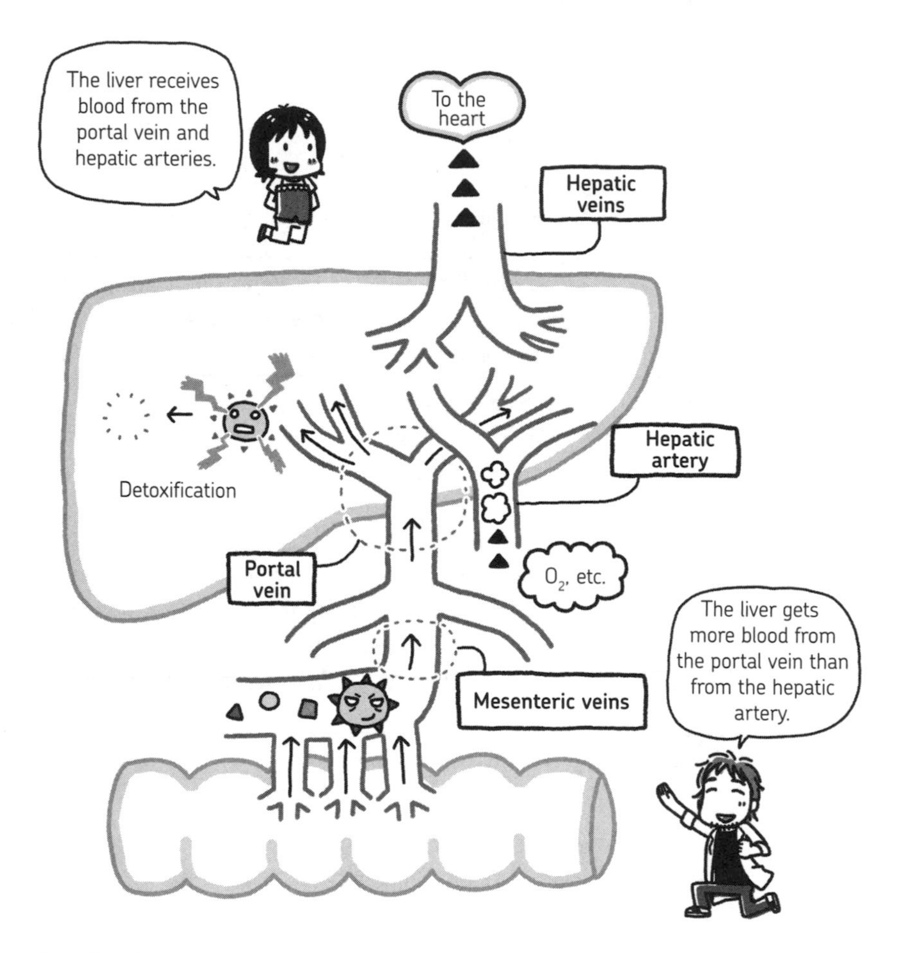

Figure 3-4: Two of the liver's major activities are detoxification and metabolism.

One job is detoxification of alcohol or other toxic substances that have entered the body. Poisons enter the body most frequently through the mouth. They are then absorbed by the alimentary canal and sent from there to the liver through the blood vessel called the portal vein.

Another job of the liver is metabolism. Nutrients taken from the alimentary canal are used as materials to synthesize or break down proteins, cholesterol, fats, and hormones (see Figure 3-5). The liver is responsible for a vast number of other crucial functions, from maintaining proper blood clotting to producing growth hormones.

Figure 3-5: The liver releases glucose when the body needs it.

The fourth job is creating bile. The bile that is created in the liver is concentrated and stored by the gallbladder. This bile aids the action of the digestive enzymes and the absorption of fats. In other words, the liver also plays an important role in digestion. Incidentally, the yellowish color of bile comes from *bilirubin*, which is a waste product produced by the metabolism of the hemoglobin in red blood cells (see Figure 3-6; see also "Red Blood Cells" on page 112).

Just think how terrible it would be if you didn't have a liver! You couldn't metabolize absorbed nutrients and you couldn't create bile. Poisons would circulate throughout the body without being detoxified, and there would be no storehouse for nutrients that are required at critical times.

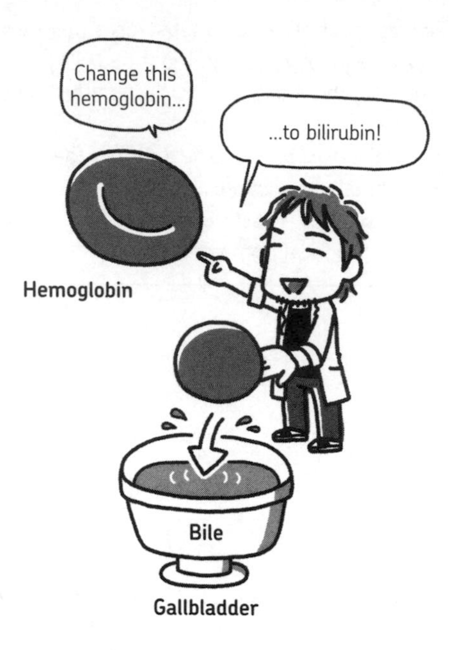

Figure 3-6: The liver breaks hemoglobin down into bilirubin to make bile, which is then stored in the gallbladder.

The liver has so many jobs. What would happen if you lost part of your liver?

The liver has an amazing ability to regenerate. Even if a doctor removes as much as three-fourths of it during an operation, it will regenerate to its original size!

4 The Kidneys and the Renal System

Cleaning Out Waste All Day, Every Day

WHAT ARE YOU DOING?

OH, YOU'RE HERE EARLY TODAY, HUH?
RIGHT NOW, I'M ORGANIZING MY COLLECTION.
INSECTS
UHHHHHHH....
LET'S SEE...
WHILE I'M TIDYING UP, I CAN TEACH YOU ABOUT THE KIDNEYS!
HOW DID THIS GET TO BE A CONVERSATION ABOUT KIDNEYS?
WHA-?

FILTERING THE BLOOD

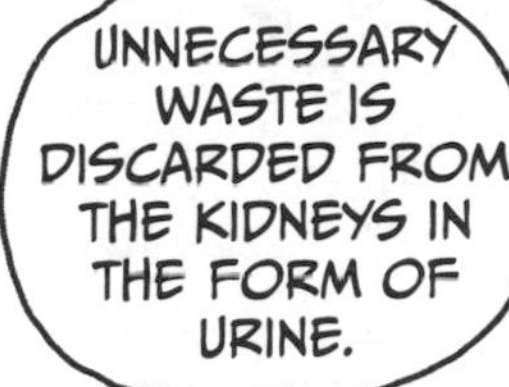

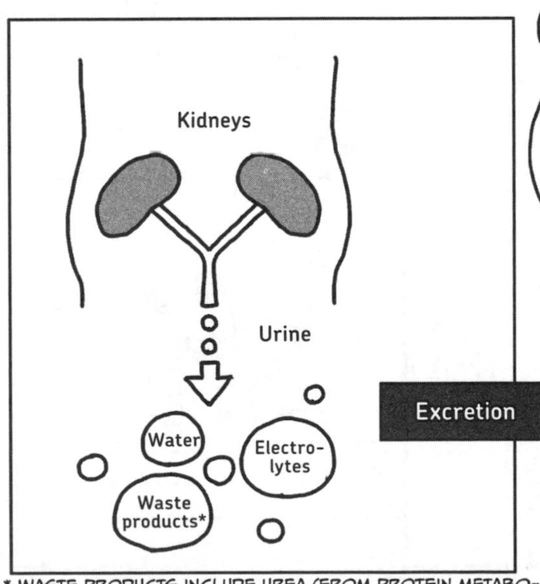

* WASTE PRODUCTS INCLUDE UREA (FROM PROTEIN METABOLISM), URIC ACID (FROM NUCLEIC ACIDS), CREATININE (FROM MUSCLE METABOLISM), UROBILINOGEN (A BYPRODUCT OF HEMOGLOBIN BREAKDOWN), AND VARIOUS PRODUCTS OF HORMONE METABOLISM.

SO WHAT DO THE KIDNEYS DO TO CREATE URINE?

えーと
FILTER BLOOD...
ER...UM...
RIGHT.

FIRST, THE KIDNEYS COARSELY FILTER THE BLOOD...
THEN THEY RECOVER ANY MATERIALS THE BODY NEEDS FROM THE FILTERED BLOOD AND DISCARD THE REST AS URINE.

HMM... LET'S TRY TO ACTUALLY DO IT.
WI-WITH THIS?

SURE, WE CAN MODEL A RENAL CORPUSCLE RIGHT HERE. THIS DESK WILL BE THE GLOMERULUS OF THE KIDNEYS.
THE GLOMERULUS IS MADE OF INTERCONNECTED CAPILLARIES.
?
バン
BAM

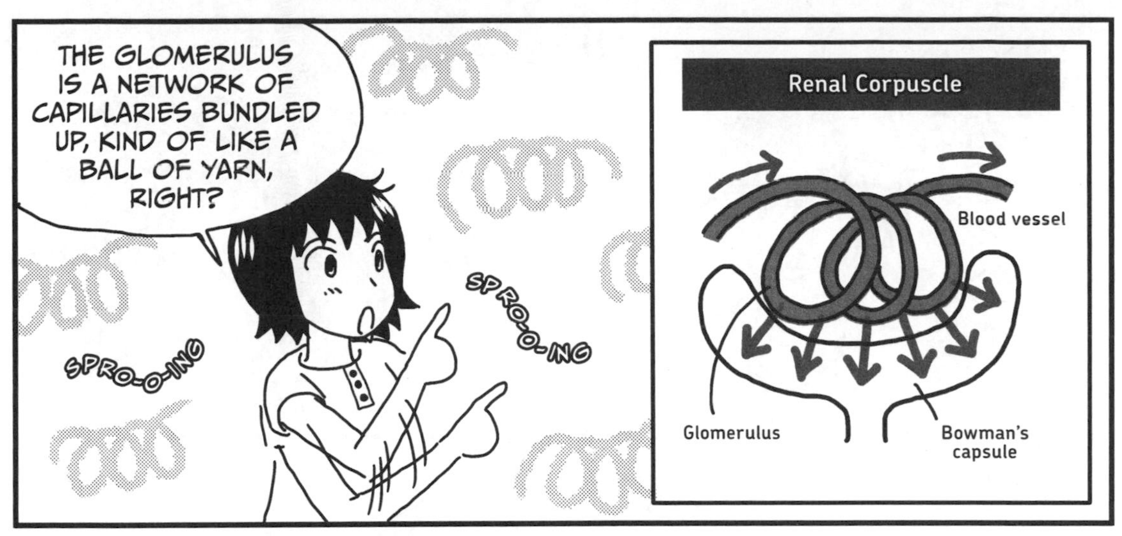
THE GLOMERULUS IS A NETWORK OF CAPILLARIES BUNDLED UP, KIND OF LIKE A BALL OF YARN, RIGHT?
SPRO-O-ING
SPRO-O-ING
Renal Corpuscle
Blood vessel
Glomerulus
Bowman's capsule

THE GLOMERULUS ACTS LIKE A SIEVE, FILTERING BLOOD AS IT PASSES THROUGH TINY OPENINGS IN THE WALLS.
Filtration
YEAH, THIS IS BOWMAN'S CAPSULE, WHICH CAPTURES THE FILTERED BLOOD.
THE STUFF PILED HERE CORRESPONDS TO THE FILTERED BYPRODUCTS OF THE BLOOD.
どっさり
Insect collection
TA-DA!
RIGHT.
BUT THIS IS JUST THE STUFF THAT WAS MECHANICALLY FILTERED BY THE FORCE OF THE BLOOD PRESSURE...

* SEE "WHAT'S IN BLOOD?" ON PAGE 111.

REABSORBING WATER AND NUTRIENTS

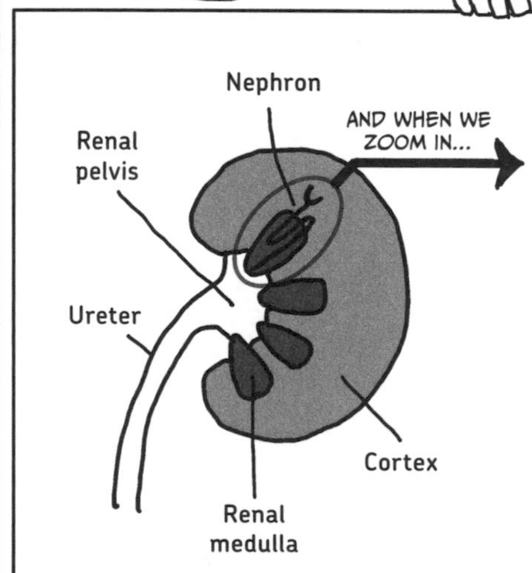

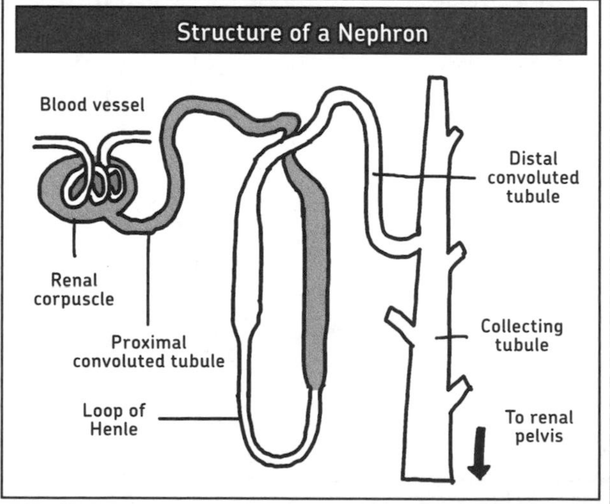

* A SINGLE KIDNEY CONTAINS APPROXIMATELY 1,000,000 NEPHRONS.

* SEE "URINE AND HOMEOSTASIS IN THE BODY" ON PAGE 92.

SO WE SHOULD DISCARD THE REMAINING STUFF IN THE RENAL TUBULE?
GARBAGE!
GARBAGE!
HOLD ON!
WAIT A MINUTE!
DISPOSING OF URINE
THAT'S NOT GARBAGE!
HUH?
STOP!
DO YOU KNOW APPROXIMATELY HOW MUCH LIQUID FROM THE GLOMERULAR FILTRATE IS DISPOSED OF AS URINE?
MAYBE... HALF...?
NOT EVEN CLOSE!!
THE AMOUNT OF WATER EXCRETED AS URINE IS 1/100 OF THE GLOMERULAR FILTRATE.
THAT MEANS 99 PERCENT IS REABSORBED BY THE RENAL TUBULE.
ONLY 1 PERCENT IS DISCARDED?!

どっさり!!
THAT'S RIGHT!
Reabsorption
ALMOST 99 PERCENT
SO MUCH STUFF
THE WORK OF MR. KIDNEY
THE NAME'S KIDNEY.
I make urine by concentrating primary urine (glomerular filtrate).
SO WHILE THE AMOUNT OF URINE PRODUCED PER DAY IS APPROXIMATELY 1 TO 2 LITERS, THE AMOUNT OF GLOMERULAR FILTRATE PER DAY IS AROUND 180 LITERS, BELIEVE IT OR NOT.
まあっ
OH MY!
WHAP
パシッ
SWEAT
SWEAT
180 L of glomerular filtrate per day
SPLASH
Renal tubule
The kidneys continue to work all the time . . .
. . . producing 1 to 2 liters of urine daily!
HUSTLE
HUSTLE
Water
Minerals
Amino acids
Vitamins
Waste
Reabsorbed!
MR. KIDNEY IS SUCH A HARD WORKER!!
SO THAT'S HOW WE LEARN ABOUT PHYSIOLOGY BY SORTING THROUGH MY GIANT PILE OF STUFF.
TA-DA!!!
えっ
へん
I UNDERSTAND WHAT THE KIDNEYS DO BUT...
REABSORPTION
TOGETHER FOREVER
I DON'T THINK I'LL EVER UNDERSTAND THIS!!

EVEN MORE ABOUT THE KIDNEYS!

Besides water, urine contains components such as sodium and other minerals, urea, uric acid, and creatinine. In a healthy person, urine is pale yellow and transparent since it contains no proteins or sugars. However, the properties of urine are not always constant as urine has a close relationship with homeostasis.

URINE AND HOMEOSTASIS IN THE BODY

The color and odor of urine can change quite a lot. After I run a marathon, the color of my urine gets really dark, but when I drink a lot of water, I urinate a large amount of almost colorless urine.

That's right. That's because the environment inside the body—such as the amount of water or pH—is being kept in a stable, constant condition.

To keep the body in this stable condition, the kidneys often have to eliminate different amounts and concentrations of material. The food and drink that we consume, the amount of activity we engage in, and how much we perspire all vary from day to day, and the substances that are disposed of as urine vary accordingly.

If you don't drink much water or if you release a lot of liquid as perspiration, you will urinate a smaller volume of more concentrated, darker urine because your body will want to dispose of as little liquid as possible. If you drink a lot of water, you will urinate a greater volume of paler urine because your body will steadily dispose of that excess water as urine.

A healthy adult produces about 1 to 2 liters of urine a day, which can be about 1 milliliter of urine per minute. If urine were discharged as fast as it's produced, we'd have to wear diapers all day! Instead, the bladder stores up urine until it starts to get full, and that's when you have to urinate. Let's look at what happens to urine in the body. Urine is produced in the kidney, and then it passes through the ureter and accumulates in the bladder. When you are standing or sitting, urine naturally falls into the bladder because of gravity. But urine is transported to the bladder even if your body is horizontal or if you are an astronaut in a state of weightlessness. This is because the ureter performs peristaltic motion to send urine to the bladder. We saw peristaltic motion earlier in the alimentary canal, too.

So how is the volume of urine in the body regulated?

The regulation of urine volume is mainly affected by two hormones. One is called the *antidiuretic hormone (ADH)*, and is secreted from the posterior pituitary gland (see "Main Endocrine Organs and Hormones" on page 221). It is also known as vasopressin. The other hormone is called *aldosterone*, which is secreted from the adrenal cortex (see "The Adrenal Glands" on page 215).

ADH is primarily secreted when blood volume is low and blood is more concentrated (see "Osmotic Pressure" on page 106 for more on blood volume), such as when the body is dehydrated. It stimulates the reabsorption of water through the renal collecting duct. This increases the water volume in the blood so that the urine is concentrated and the volume decreases.

Secretion of aldosterone is similarly triggered by decreases in blood volume and even blood pressure. Aldosterone stimulates the reabsorption of sodium from the renal tubule into the bloodstream. Through osmosis, water follows the sodium and is reabsorbed into the bloodstream (see Figure 4-1), reducing the urine volume as a result.

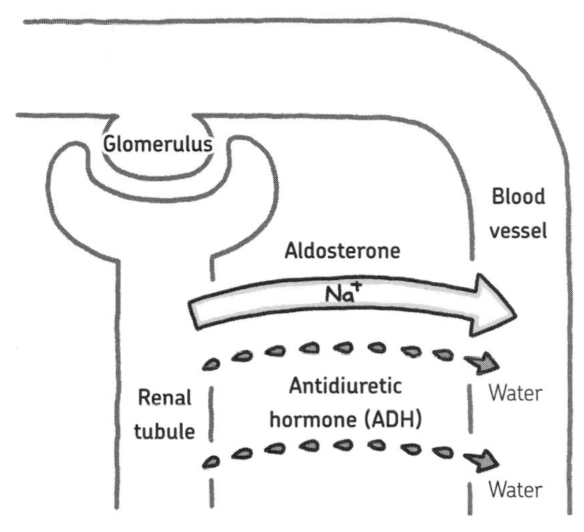

Figure 4-1: Water reabsorption from urine

> **Homeostasis and the Body's Crisis Management System**
>
> Living creatures must maintain homeostasis despite changes inside and outside the body. Homeostasis is an active and dynamic process: the body regulates its temperature and pH, fights off invading pathogens, and heals wounds. Many different systems, from the autonomic system to the endocrine system, work together to maintain a delicate equilibrium. In addition to the many bodily functions that go unnoticed, the drive for homeostasis can affect the way we feel and act. The body becomes hungry when blood sugar drops and thirsty when dehydrated, for example. If the environment inside the body gets out of balance, the body will attempt to return to its normal state.

THE PROCESS OF URINATION

Let's go over what happens during urination. When there's room in the bladder to store more urine, the smooth muscles in the wall of the bladder relax while both the internal sphincter (smooth muscles) and external sphincter (skeletal muscles) located at the exit of the bladder contract, holding the urine in. When approximately 200 to 300 milliliters of urine have accumulated, the bladder walls stretch out and a message is sent to the brain, triggering the urge to urinate. When you rush to the bathroom to pee, the muscles in the bladder walls contract, the internal and external sphincters relax, and you urinate (see Figure 4-2).

While urine accumulates, the bladder is stretched. The smooth muscles in the walls of the bladder relax, and both the internal and external sphincter contract.

When you want to urinate, the muscles in the walls of the bladder contract, and the internal and external sphincters open to release urine.

Figure 4-2: Bladder muscles controlling the release of urine

Usually, once urination begins, it continues until all urine is expelled from the bladder. However, if a problem occurs, some urine can remain inside the bladder. This is called *residual urine*, and it can cause an increased risk of infection or even kidney dysfunction.

Approximately how much urine can the bladder hold?

Normally, when approximately 200 to 300 milliliters of urine has accumulated, you get the urge to urinate. However, if you really hold it in, you can store up to 500 milliliters. In fact, under some circumstances, the bladder is said to be able to hold up to *800 to 1000 milliliters*!

The Urinary Tract

Together the kidneys, ureter, bladder, and urethra make up the urinary tract. The bladder and internal sphincter are smooth muscles (involuntary muscles), and the external sphincter is a skeletal muscle (voluntary muscle). Therefore, urination is a complex, high-level operation in which involuntary actions and voluntary actions are intertwined.

The urethra is different in males and females, isn't it?

That's right. The male urethra is 16 to 18 centimeters (6 to 7 inches) long, while the female urethra is only 3 to 4 centimeters (1 to 1.5 inches) long.

This is one of the reasons why females are more susceptible to cystitis or a urinary tract infection (UTI). Bacteria can enter through the urethral orifice, reach the bladder, and cause infection (see Figure 4-3).

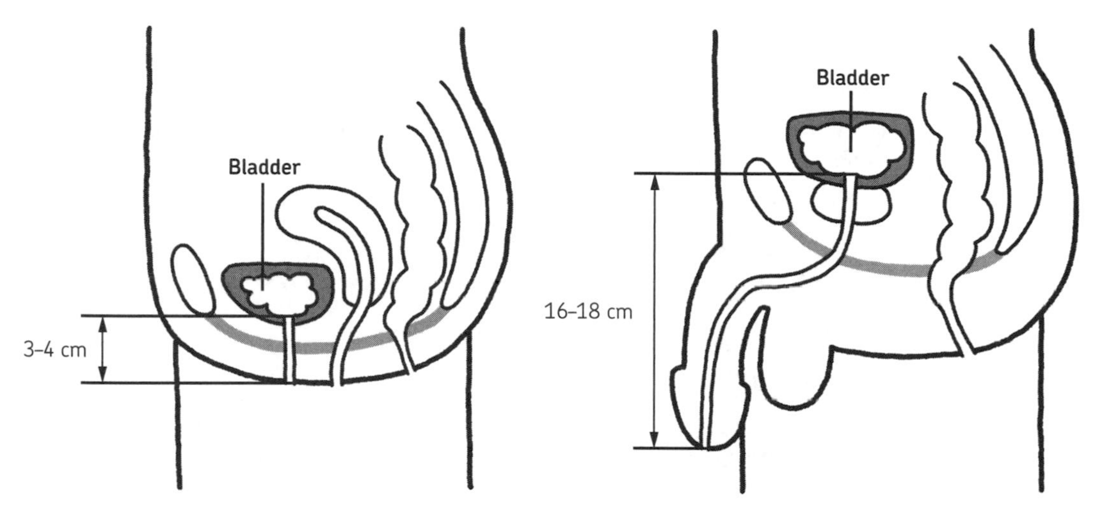

Figure 4-3: Different lengths of male and female urethras

MONITORING BLOOD IN THE KIDNEYS

The kidneys are also endocrine organs that secrete hormones.

What? They don't just create urine?

No, the kidneys secrete hormones related to blood pressure and the creation of blood. Since blood is filtered in the kidneys, a large volume of blood is always passing through them. They monitor the blood flowing in, and if they discover a problem, they secrete a hormone to resolve that problem. It's a well-balanced system, isn't it?

KIDNEY FUN FACTS!

The two kidneys are located on the left and right sides of the body, at the lower back. When functioning normally, a single kidney is enough to work for the entire body! That means a patient suffering from end-stage renal disease may be eligible to receive a kidney from a sibling, and both of the siblings can then live with a single kidney.

The kidneys carefully monitor two factors in particular: blood pressure and oxygen concentration (see Figure 14-4). If blood pressure drops, the blood can no longer be filtered properly in the renal corpuscle. When this happens, the kidneys secrete the hormone renin, which then influences the hormones angiotensin and aldosterone, which in turn raise the blood pressure.

Figure 4-4: The kidneys secrete hormones to regulate blood pressure and oxygen concentration.

Low oxygen concentration of the blood flowing into the kidneys indicates that there are not enough red blood cells carrying oxygen. The kidneys will then secrete the hormone erythropoietin to influence the creation of more red blood cells in the bone marrow.

One more job that the kidneys perform is vitamin D activation, which allows us to metabolize calcium and strengthen our bones. Vitamin D can be absorbed from the foods we eat or created in the skin when the skin is exposed to the sun. However, it cannot strengthen bones directly. First, it must be converted by the kidneys to the active form of vitamin D, a substance called calcitriol, before the body can metabolize calcium (see Figure 4-5).

Figure 4-5: The kidneys convert vitamin D into calcitriol, which is necessary for the metabolism of calcium.

WHEN THE KIDNEYS STOP WORKING

The kidneys eliminate waste products that are produced in the body, as well as excess liquids and minerals. The average amount of urine produced in a day for a healthy adult is between 1 and 1.5 liters. However, this volume can change depending on how much water is drunk or how much the body has perspired, so it can be less than 1 liter or as much as 2 liters.

The production of less than 400 milliliters of urine per day is called *oliguria*. Oliguria can cause serious health complications because at least 400 milliliters of urine are required to dispose of the waste products produced in the body in a day. *Anuria* occurs if the daily urine volume is below 50 milliliters.

What happens when the kidneys stop working?

That is called *renal insufficiency* or *renal failure* (see Figure 4-6). There are various degrees of severity, ranging from mild cases to conditions in which kidney function is almost completely lost.

Renal insufficiency occurs when the kidneys can't perform their functions because they are diseased, infected, or overloaded with toxins or because their blood supply is interrupted due to injury. When renal insufficiency occurs, water, acid, potassium, and waste products are not disposed of effectively and can accumulate in the body. This is similar to what happens when a swimming pool filter breaks down and the water steadily gets dirtier. If waste products are not eliminated, the ensuing buildup of toxins can lead to *uremia*, a serious and even deadly illness.

Figure 4-6: Renal insufficiency

Renal insufficiency can also mean that excess water is not disposed of, in which case the blood volume will increase, the heart will have to work harder, and heart failure may occur. Heart failure can in turn cause the lungs to become flooded—a condition called *pulmonary edema*, which can lead to respiratory failure. Another result of renal insufficiency is a buildup of acid (*acidosis*), since acid is not eliminated. If excessive potassium accumulates in the body, the heart muscles may convulse irregularly (*ventricular fibrillation*), which can lead to sudden death.

Kidney Problems and Dialysis

Since the kidneys are also involved in the regulation of blood pressure, the production of blood, and the metabolism of calcium, problems in the kidneys can cause high blood pressure, anemia, or bone fractures, as well as a buildup of toxins and water. For patients with severe kidney injury, machines have been developed to remove waste products and excess water from the body. This process is called *dialysis*.

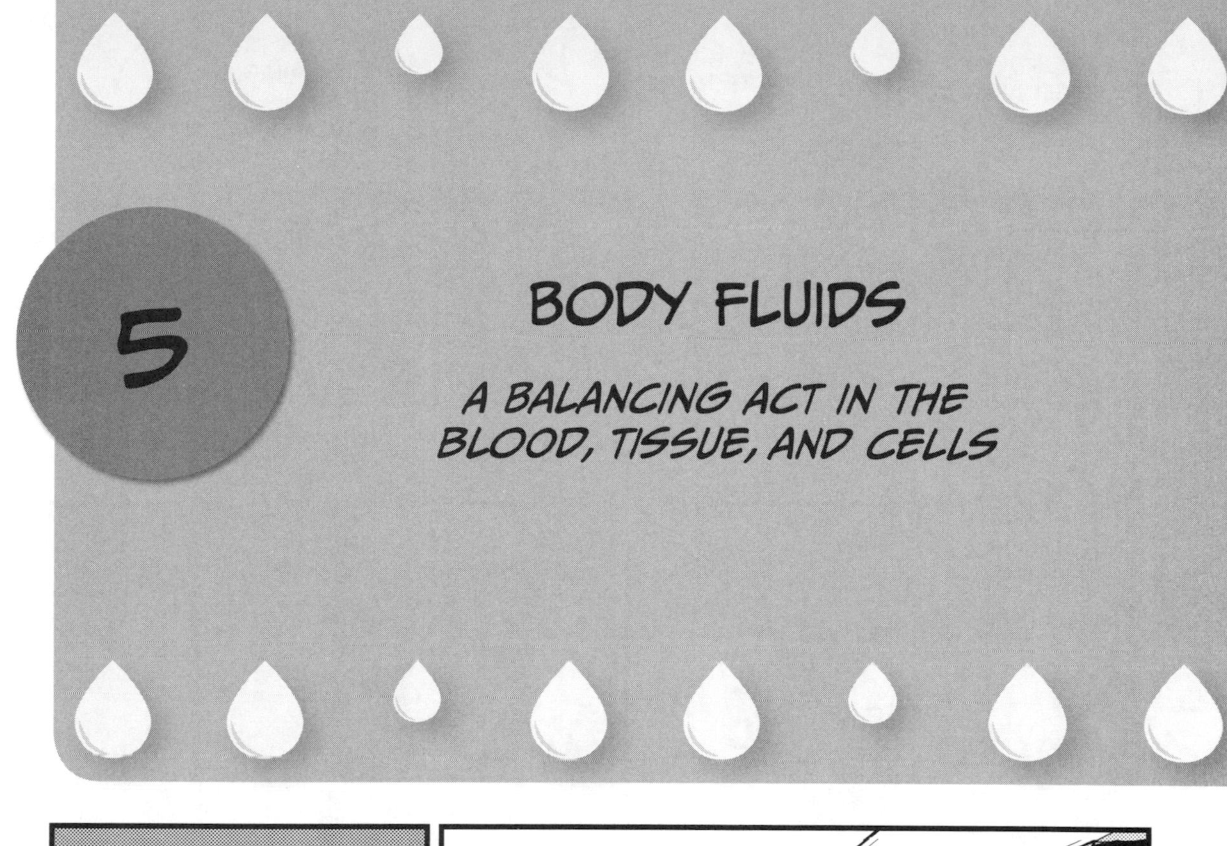

5 Body Fluids

A Balancing Act in the Blood, Tissue, and Cells

EVERYONE'S SO EXCITED ABOUT THIS MARATHON.
IT'S TOO MUCH PRESSURE...
KOUJO Medical School
Marathon Race
THEY'RE ALL COMING BACK EARLY FROM SUMMER VACATION JUST TO WATCH!
GO KUMIKO!
THIS BOOK IS MONSTROUS!
PHYSIOLOGY
THUMP!
I NEED TO TRAIN, BUT I HAVE TO SPEND ALL DAY READING THIS!!
3 in
WHO COULD GET THROUGH SUCH A THICK, BORING TEXTBOOK??
PHYSIOLOGY
PHYSIO
BUT THANKS TO PROFESSOR KAISEI, PHYSIOLOGY IS STARTING TO GET MORE INTERESTING...
WHAT ARE YOU MUTTERING ABOUT?
HAVE YOU BEEN STUDYING FOR YOUR MAKEUP EXAM?!
AH...
YES?

EX-EX-EX-
EXCUSE ME!!!
ピュー
WHOOSH
HMMPH. SHE CAN RUN ALL SHE LIKES...
BUT SHE NEEDS TO PASS THIS TEST.
HUMANS ARE 60 PERCENT WATER
103
DEPARTMENT OF SPORTS AND HEALTH SCIENCE
HEALTH SCIENCE CLASSROOM
YEAH, IT WAS AN ABSOLUTE DISASTER...
I BROKE OUT INTO A COLD SWEAT AND SPRINTED ALL THE WAY HERE. NOW I'M SO DEHYDRATED I FEEL DIZZY!
AHAHAHAHA
HA HA HA...
SO THEN YOU CAME RACING OVER HERE?
THAT SOUNDS AWFUL.
AH, THIS IS PERFECT FOR YOU.
ELECTRO! LITE
DRINK THIS.
IT'S A SPORTS DRINK THEY WERE GIVING AWAY LAST WEEK.
THANKS A LOT.

GLUG GLUG
BETTER DRINK UP!
YOU'LL NEED IT. ABOUT 60 PERCENT OF THE HUMAN BODY IS MADE OF WATER.
COOL, RIGHT?

I KNEW THAT! TWO THIRDS OF THE WATER IN OUR BODIES IS INSIDE OUR CELLS (INTRACELLULAR FLUID) AND THE REST IS OUTSIDE (EXTRACELLULAR FLUID).
OF THE EXTRACELLULAR FLUID, THREE QUARTERS IS INTERSTITIAL FLUID, INSIDE THE MUSCLE TISSUE (ALSO CALLED TISSUE FLUID), AND THE REST IS EITHER PLASMA INSIDE OUR BLOOD (INTRAVASCULAR FLUID) OR FLUID IN THE BODY CAVITY!
Extra-cellular fluid
Intracellular fluid
Intra-vascular fluid
Interstitial fluid
OH...THAT'S AWESOME!

WAS ALL THAT WRITTEN ON THE LABEL?
I GUESS YOU CAUGHT ME.
SO...
CARE TO LEARN MORE ABOUT FLUIDS IN THE BODY?
YEAH, SOUNDS GREAT!

OKAY.
START BY IMAGINING A RICE PADDY.
Rice paddy

MS. KARADA...
DO YOU KNOW HOW IRRIGATION OF A RICE PADDY WORKS?
WHA-WHAT? RICE PADDIES...
?

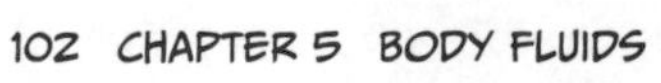

THE JAPANESE COUNTRYSIDE IS FILLED WITH RICE PADDIES.
KUMIKO, ARE YOU STUDYING HARD?
LEAVE ME ALONE!
SIGH...
RICE PADDIES ARE REALLY GREAT.
THERE ARE INSECTS EVERYWHERE...
AHEM...
ANYWAY, WE CAN LIKEN THE IRRIGATION IN PADDY FIELDS TO HOW FLUIDS WORK INSIDE YOUR BODY.
THE RICE PLANTS ARE LIKE CELLS...
THE IRRIGATED PADDY FIELDS ARE LIKE THE INTERSTITIAL FLUID BETWEEN CELLS...
AND THE CHANNELS DRAWING WATER FROM THE STREAM TO THE IRRIGATED PADDY FIELDS ARE LIKE THE BLOOD VESSELS.
WATER AND NUTRIENTS ARE CARRIED BY THE CHANNELS.
Blood
THIS SUPPLIES INTERSTITIAL FLUID BETWEEN CELLS.
Interstitial fluid
THE CELLS ABSORB WHAT THEY NEED FROM THAT FLUID!
Cells
I SEE...
THIS IS JUST WEIRD ENOUGH TO MAKE SENSE!

INTERSTITIAL FLUID IS THE FLUID OUTSIDE OUR CELLS. ITS COMPONENTS ARE VERY SIMILAR TO THOSE OF PLASMA.
Extracellular fluid
Intracellular fluid
INTERSTITIAL FLUID AND BLOOD ARE CALLED EXTRACELLULAR FLUID SINCE THEY ARE OUTSIDE OF THE CELLS.
AND THE FLUID INSIDE CELLS IS CALLED INTRACELLULAR FLUID.
THE COMPONENTS OF EXTRACELLULAR FLUID AND INTRACELLULAR FLUID ARE DIFFERENT, RIGHT?
HOW ARE THEY DIFFERENT?
RIGHT.
BUT BEFORE WE TALK ABOUT HOW THEY'RE DIFFERENT, LET'S TALK ABOUT WHY THEY'RE DIFFERENT.
LIFE BEGAN IN THE PREHISTORIC OCEANS.
INTRACELLULAR FLUID WAS ENCLOSED BY CELL MEMBRANES AND SEPARATED FROM THE SURROUNDING OCEAN WATER, WHICH YOU MIGHT THINK OF AS EXTRACELLULAR FLUID.
Ocean water
Intracellular fluid
Cell membrane
EXTRACELLULAR FLUID IN LARGER, MORE COMPLEX ORGANISMS HAS MANY OF THE SAME COMPONENTS AS OCEAN WATER, PROVIDING AN ENVIRONMENT FOR CELLS THAT IS SIMILAR TO THE OCEAN IN PREHISTORIC TIMES.
Extracellular fluid
Intracellular fluid
AS INDIVIDUAL CELLS JOINED TOGETHER AND MULTICELLULAR ORGANISMS EVOLVED, EVENTUALLY, THEY EMERGED FROM THE OCEAN ONTO DRY LAND. BUT IN A WAY, IT'S AS IF THE CELLS REMAINED IN THEIR OWN INDIVIDUAL OCEAN (EXTRACELLULAR FLUID).
Extracellular fluid
Cell
Intracellular fluid
Extracellular fluid
Blood*
* BLOOD CAN ALSO BE THOUGHT OF AS EXTRACELLULAR FLUID THAT CARRIES OXYGEN AND NUTRIENTS TO EACH CELL.

HMM...
SO EXTRACELLULAR FLUID IS A LOT LIKE THE OCEAN... AND THE OCEAN IS VERY SALTY...
Na+
Na+
Na+
EXTRACELLULAR FLUID IS SALTY TOO, ISN'T IT?
Na+!
Na+
K+
EXTRACELLULAR FLUID HAS A LOT OF SODIUM (Na+).
BUT INSIDE CELLS IS A LOT OF POTASSIUM (K+) BUT NOT MUCH SODIUM (Na+).
K+
Na+
K+
Na+
VERY GOOD!
CLAP CLAP
K+
Na+
THE WATER AND NUTRIENTS FROM THE IRRIGATION CHANNELS (BLOOD) FEED INTO THE WATER IN THE PADDY FIELDS (INTERSTITIAL TISSUE) AND THEN INTO THE RICE PLANTS (CELLS).
BODY FLUIDS ARE REALLY IMPORTANT, AREN'T THEY?
THAT'S RIGHT! IN OTHER WORDS, THE NUTRIENTS AND LIQUID IN THE BLOOD ARE TRANSFERRED TO THE CELLS THROUGHOUT THE ENTIRE BODY THROUGH THE INTERSTITIAL TISSUE.
YOUR BODY FLUIDS MUST STAY IN A DELICATE BALANCE TO KEEP YOUR CELLS FUNCTIONING PROPERLY.

OSMOTIC PRESSURE

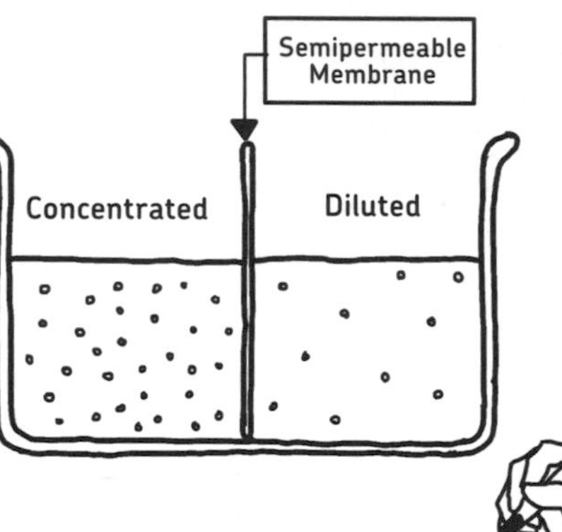

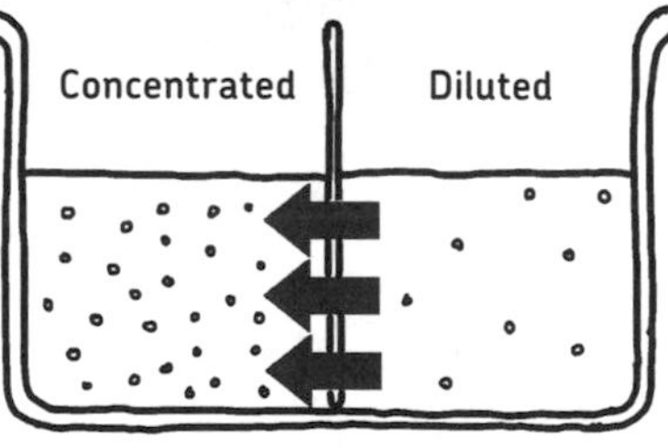

ACTUALLY, THEY'RE A BIT DIFFERENT. DIFFUSION IS THE PHENOMENON IN WHICH PARTICLES SPREAD OUT FROM A HIGHLY CONCENTRATED AREA TO A LESS CONCENTRATED AREA. IN OSMOSIS THE PARTICLES *MOVE* FROM A LOW CONCENTRATION TO A HIGH CONCENTRATION.

Osmosis is the movement of *water* particles between two volumes of liquid (separated by a semipermeable membrane) with different concentrations of a solute. The movement of water is *from low concentration to high concentration.*

Diffusion is the movement of *solute* particles as they spread out. The movement of particles is *from high concentration to low concentration.*

* AN *ELECTROLYTE* IS A SUBSTANCE THAT IS DIVIDED INTO POSITIVE AND NEGATIVE IONS WHEN DISSOLVED IN A LIQUID SUCH AS WATER AND THEREFORE CARRIES AN ELECTRICAL CHARGE. SODIUM AND POTASSIUM ARE BOTH EXAMPLES OF ELECTROLYTES.

FOR BODY FLUIDS, OSMOTIC PRESSURE IS PARTICULARLY AFFECTED BY THE CONCENTRATION OF PROTEINS, SPECIFICALLY ALBUMIN.
I'M A PROTEIN!
Albumin
OSMOTIC PRESSURE CAUSED BY ALBUMIN IS CALLED ONCOTIC PRESSURE OR COLLOID OSMOTIC PRESSURE.
Oncotic Pressure or Colloid Osmotic Pressure
SO ALBUMIN IN THE BLOOD CREATES ONCOTIC PRESSURE, WHICH DRAWS FLUID INTO THE BLOODSTREAM FROM THE SURROUNDING TISSUE?
THAT'S RIGHT!
IF AN ILLNESS REDUCES THE AMOUNT OF PROTEINS IN THE BLOOD (HYPOPROTEINEMIA), THE ONCOTIC PRESSURE OF BLOOD WILL BE LOWER, AND THERE WILL BE LESS FORCE DRAWING MOISTURE FROM THE INTERSTITIAL FLUID INTO THE BLOODSTREAM. THIS CAUSES INTERSTITIAL FLUID TO BUILD UP (CALLED *EDEMA*).
THERE'S TOO MUCH WATER TO BRING IN!
WE CAN'T KEEP UP.
H_2O
Serum Albumin
Interstitial Fluid
Semipermeable Membrane
Blood
A-HA!
WHEN I OVERSLEEP AND MY FACE GETS PUFFY, THAT MUST BE HYPOPROTEINEMIA, RIGHT?
THAT'S JUST OVER-SLEEPING...
ラー

EVEN MORE ABOUT BODY FLUIDS AND THE BLOOD!

Water is essential in all the normal functions of the human body. You need water to circulate blood throughout the body, maintain body temperature, dispose of unneeded waste products, and secrete digestive fluids.

KEEPING HYDRATED

Approximately 60 percent of human body weight is made up of water. Your body needs to maintain that proportion, so liquids that are expelled from the body need to be replaced (see Figure 5-1). Liquids are taken into the body from both water and other drinks, and from particular foods, like vegetables, meat, and grains. An adult male cycles around 2600 milliliters per day: that's 2.6 liters going in and 2.6 liters going out!

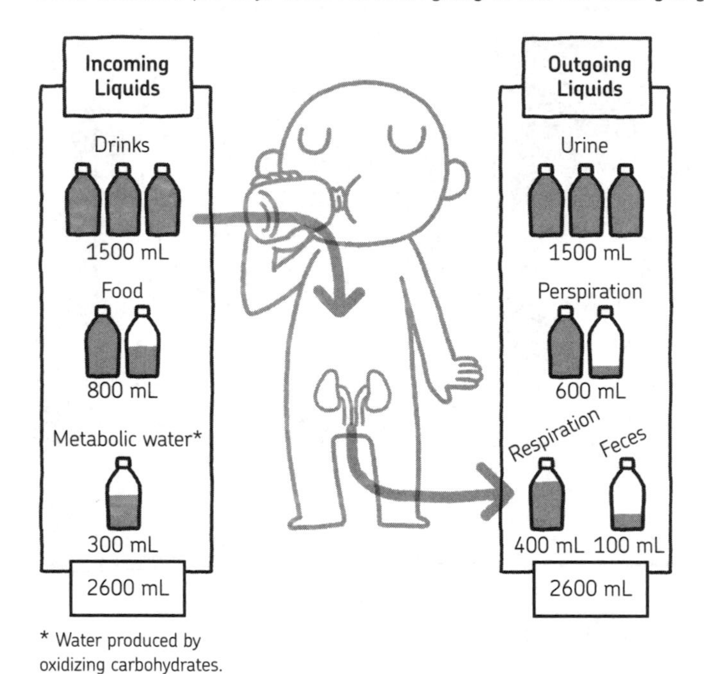

Figure 5-1: Average volume of liquid entering and exiting body per day (for an adult male)

In a healthy person, the amount of liquid entering and exiting the body are the same, aren't they?

That's right. If the amount of liquid output is too little, edema occurs, and if it is too much, dehydration occurs.

What happens when the body is dehydrated?

Dehydration can cause various problems. Circulatory failure, impaired consciousness, and an increase in body temperature can all result from serious dehydration. Dehydration can even lead to death.

Dehydration can occur in several ways. *Intracellular dehydration* is mainly caused by excessive perspiration or an insufficient intake of liquids. As this occurs, the concentration of solutes in the extracellular fluid becomes greater, increasing the extracellular osmolarity. This in turn causes water to move from the intracellular space to the extracellular space, and the person feels thirsty as a result. When the concentration of extracellular fluid is too high, it's called hypertonicity.

On the other hand, *extracellular dehydration* occurs when the circulating blood volume decreases, causing blood pressure to drop significantly. This can be caused by a deficiency of sodium in the bloodstream; sodium helps retain liquid in intracellular fluids (especially blood plasma). If pure water were infused into the bloodstream in an intravenous drip, it would be too hypotonic for the body (that is, it wouldn't have enough solutes, like salt and other electrolytes), and cells might swell and die as result. In a drip, electrolytes must be added to the water in order for the body to safely tolerate it.

WHO'S AT RISK?

Since liquids account for a greater proportion of the body weight of infants, and babies lose more liquids to perspiration and respiration than adults or older children do, infants can easily suffer dehydration. Elderly people with a reduced sense of thirst or less ability to conserve water are also at risk of dehydrating.

WHAT'S IN BLOOD?

So far, we've been talking about water and fluids in the body in general. It's time we start talking in detail about one body fluid in particular: blood. Blood has many important characteristics and plays many vital roles in the body.

Let's start by thinking about how you would draw and examine a blood sample in a lab. When blood is drawn, an anticoagulant is added and the test tube is centrifuged. The blood cells will sink to the bottom, and a clear liquid will rise to the top, as shown in Figure 5-2. (When a liquid settles on top of another liquid or a solid, the upper layer is referred to as *supernatant*.)

This clear liquid at the top is *blood plasma*. Blood plasma is the extracellular fluid that carries blood cells throughout the body and helps remove waste products. It is mostly made up of water but also contains essential proteins like antibodies and enzymes.

The blood cells at the bottom of the tube can be divided into three general categories: red blood cells, white blood cells, and platelets. The majority of blood cells are red blood cells, which is why blood appears red. A test that finds the percentages of the various cells in the blood is called a *hematocrit*. I'll now introduce each of these kinds of blood cells one by one.

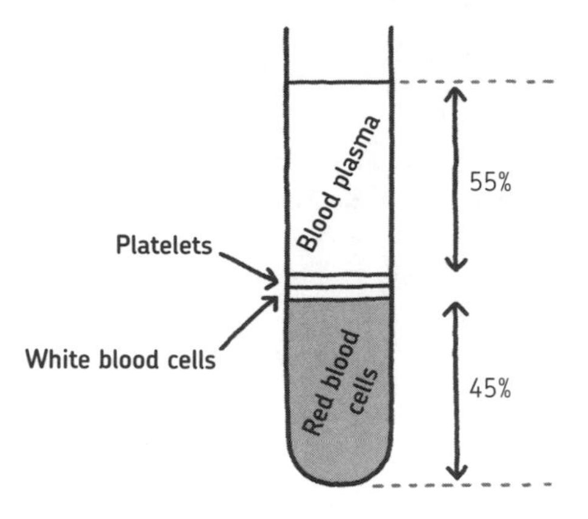

Figure 5-2: The components of blood and their distribution

RED BLOOD CELLS

The most common type of blood cell is the *red blood cell*. Red blood cells are created in bone marrow. They don't have a cell nucleus, meaning they can't subdivide to make more cells, and they have a concave disk shape. This shape is advantageous because it increases the cell's surface area, allowing it to bond to more oxygen molecules. Red blood cells can also elongate to enter a capillary that's thinner than the blood cell's usual diameter, as shown in Figure 5-3.

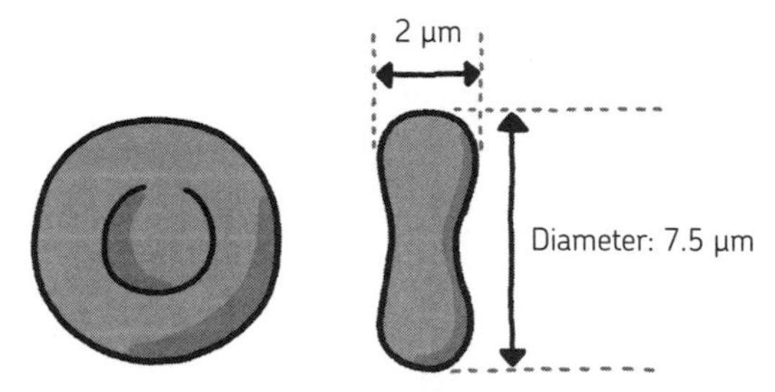

1 μm (micrometer): 1/1000 of 1 millimeter

Figure 5-3: How big is a red blood cell?

Red blood cells transport oxygen using *hemoglobin*. Hemoglobin is a combination of a pigment called heme, which contains iron, and a protein called globin. Hemoglobin bonds easily to oxygen, which it picks up in the alveoli of the lungs. When hemoglobin picks up oxygen, it becomes a bright red color (see Figure 5-4). This is why arterial blood is bright red, while peripheral venous blood (blood whose oxygen has been transported to other parts of the body) is dark red.

Figure 5-4: When oxygen finds hemoglobin, they immediately bond, and the hemoglobin becomes bright red.

When there's not enough iron in the blood, anemia occurs. Isn't iron-deficiency anemia more common in women than men?

That's true. Since iron is an ingredient of hemoglobin, someone who does not take in enough iron will not make enough hemoglobin, and in turn the number of red blood cells will decrease.

Because women have menstrual cycles and lose a fixed amount of blood every month, they are more likely to become anemic. Women also naturally have a lower number of red blood cells and lower hemoglobin concentration than men.

What Is Anemia?

Anemia is a condition caused by a reduction in the ability to transport oxygen due to a reduction in hemoglobin. It occurs when the hemoglobin concentration, or the number of red blood cells, has dropped below normal. *Iron-deficiency anemia* is the most common type of anemia, but there are other, more serious types of anemia, such as hemolytic anemia, caused by an abnormal breakdown of red blood cells, and aplastic anemia, caused by a disorder of the bone marrow that interferes with its production of red blood cells.

Red blood cells have a particular life span, don't they?

Yep. The life span of a red blood cell is roughly 120 days. When they approach the end of their life spans, red blood cells are broken down and recycled by the liver and spleen. Figure 5-5 shows how heme becomes bilirubin and is excreted in bile. Iron is removed from heme and stored for later use.

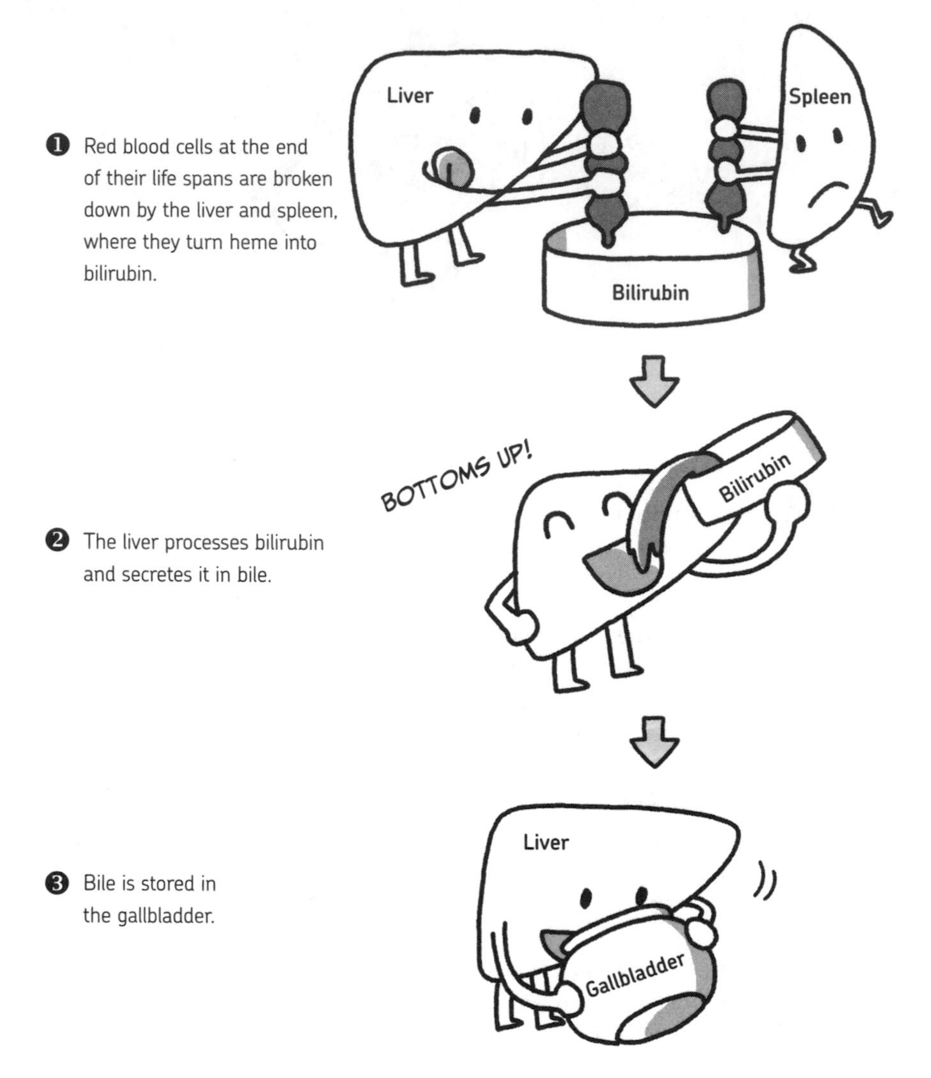

Figure 5-5: Heme being broken down into bilirubin, which is then secreted in bile.

BLOOD FUN FACTS

Blood is classified into the ABO blood types according to *antigens* in the membranes of red blood cells. The antigen for each type is like a special signature, and the immune system will attack all cells with antigens that don't match the antigens corresponding to the person's own blood type. That's why you can receive blood only from someone with the same blood type as you in a blood transfusion. The most common blood types are O and A, while AB is rarer.

WHITE BLOOD CELLS

Blood contains approximately 5000 to 8000 *white blood cells* per microliter. White blood cells act as the body's defense forces. One of their jobs is to repel foreign enemies such as invading bacteria and viruses.

White blood cells are broadly divided into granulocytes, monocytes, and lymphocytes, as shown in Figure 5-6. These are further classified into numerous types, each with different characteristics and responsibilities. All are extremely skillful at cooperating to repel invaders.

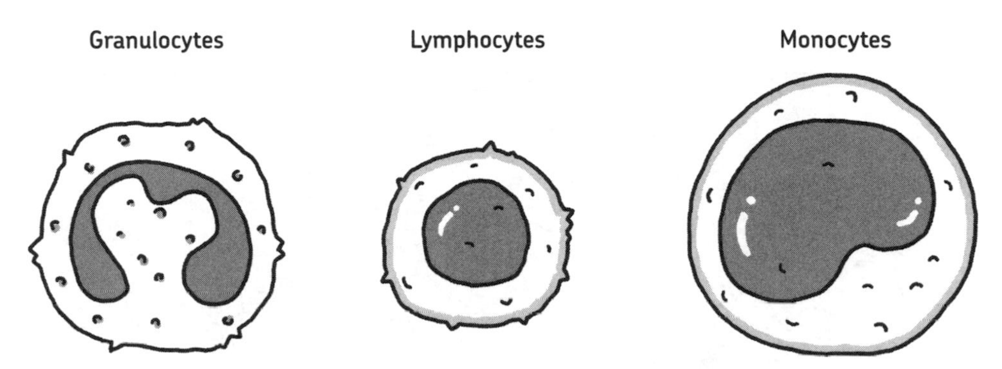

Figure 5-6: Types of white blood cells

Granulocytes are characterized by intracellular particles called *granules*. The three types of granulocytes are neutrophils, eosinophils, and basophils. The majority of granulocytes are neutrophils, which destroy any existing invaders by engulfing and devouring them. This is called *phagocytosis*. The pus that sometimes exudes from a wound partly consists of neutrophils that have performed phagocytosis and died. There are relatively few eosinophils and basophils, but these types of granulocytes contribute to phagocytosis and allergic reactions.

Lymphocytes—which include B cells, T cells, and NK cells (or natural killer cells)—are major actors in the immune system. The T cells direct the immune response, and B cells prepare and release the appropriate antibodies. Along with the NK cells, the T cells also destroy any infected cells.

Monocytes are large, round cells when they are found in blood vessels. However, when they move through the blood vessel wall into tissue, they change shape and become macrophages. *Macrophages* extend tentacles to grab and devour invaders. This is another form of phagocytosis.

Specialist B Cells

Once a foreign enemy like bacteria invades, B cells remember information about it. So if that substance invades a second time, the body can quickly identify it and release a large number of antibodies to repel it. However, a single B cell can only remember a single foreign enemy so millions of B cells exist in the human body to fight off the many potential attackers.

How do our bodies repel foreign substances?

When a foreign substance invades, neutrophils and macrophages (which can be thought of as scouting parties) rush in first and voraciously devour them. The macrophages then present the fragments of the devoured enemy to the T helper cells, (which are the "commanding officers") to say, "This is what we're facing!" Then the general offensive begins (see Figure 5-7).

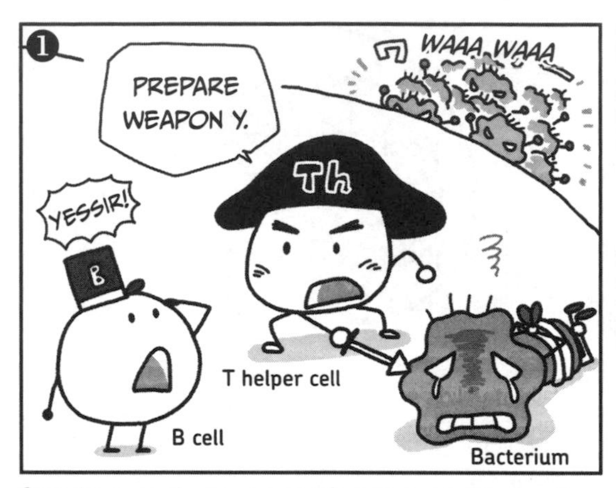

Once the macrophages present information about the invading forces to the T helper cells, the T helper cells instruct the B cells to prepare antibodies designed to fight that specific foreign substance.

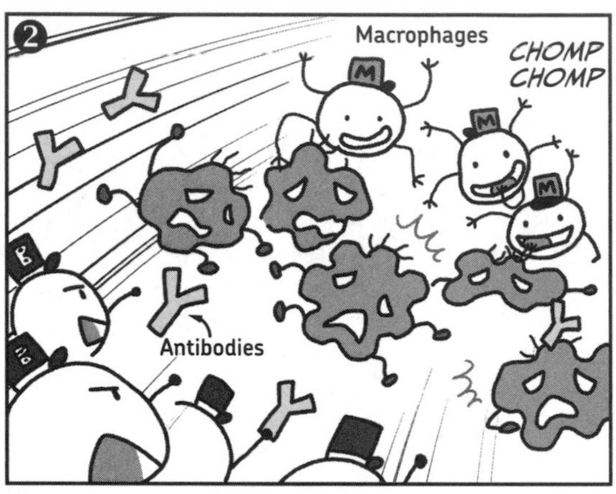

The B cells prepare the antibodies and release them into the blood. The antibodies tag and neutralize the foreign substances, and the macrophages then devour and eliminate the neutralized enemy.

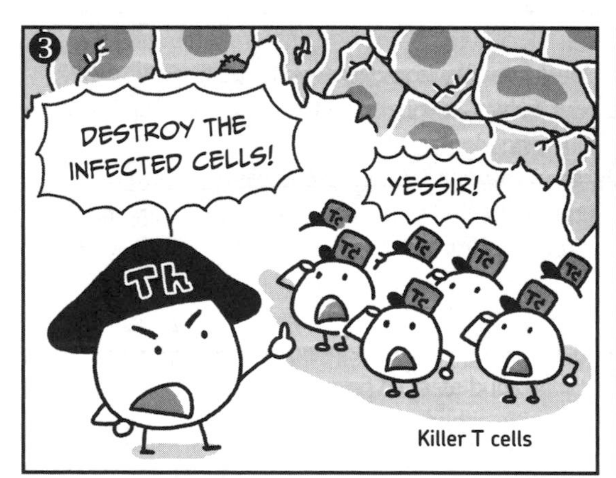

The T helper cells order the killer T cells to destroy any cells that were infected or damaged by the foreign substance.

Once the foreign substance is entirely eliminated, the suppressor T cells call a stop to the defensive response.

Figure 5-7: The white blood cell defense forces

An Allergy Is an Overreaction of the Immune System

An *allergy* can be thought of as a runaway immune function. Instead of repelling invading viruses or bacteria, the immune system is reacting to a substance that is not actually harmful, such as food or pollen.

Food and pollen are common causes of allergic reactions. Allergy to pollen, often called hay fever, results in nasal inflammation or conjunctivitis. Bronchial asthma and atopic dermatitis (also known as *eczema*) are also common allergic reactions.

The prevalence of allergies is increasing in both in adults and children. Although environmental factors (such as diet, reduced physical activity, better hygiene, or even changes in home heating and ventilation systems) are thought to be involved with the increase in these conditions, the actual reason is not yet known.

PLATELETS

A *platelet* is a cell involved in hemostasis, which is the process that stops bleeding. It has no nucleus and is produced in a bone marrow megakaryocyte (meaning *large nucleus cell*). One microliter of blood contains approximately 300,000 platelets. This might sound like a lot, but it is very few compared to the number of red blood cells. When blood is centrifuged, the platelet layer is extremely thin.

When a blood vessel is damaged and bleeding occurs, the platelets are the first to act. First, they gather at the damaged location to form a temporary plug. Then the platelets burst open and release substances that speed up hemostasis. Those substances react and eventually turn fibrinogen (a substance contained in blood plasma) in the blood into a fibrous substance called *fibrin*, which forms a mesh at the site of the wound. Red blood cells become caught in the mesh and clump together to form a robust dam to stop the bleeding. This clump is called a *blood clot* (Figure 5-8).

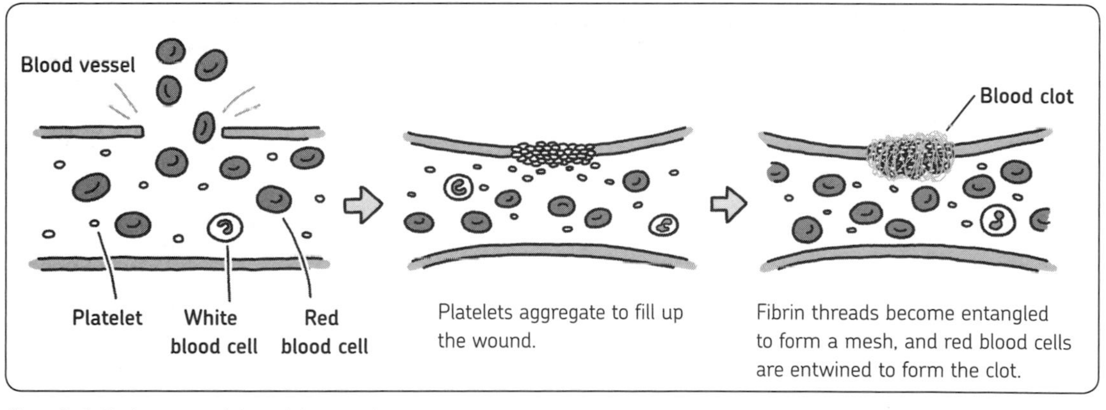

Figure 5-8: Closing a wound through hemostasis

Applying pressure is an effective way to stop bleeding, isn't it?

Yes, pressure constricts the capillaries and blood vessels, slowing down the bleeding. This gives the blood time to clot. You can sometimes stop bleeding from capillaries or slender veins just by applying pressure.

Blood that's been drawn naturally solidifies just like blood that's released by bleeding. The solidification of blood is called *coagulation*. During an examination of drawn blood, chemicals are often mixed into the blood to prevent it from solidifying. Blood coagulates because blood plasma itself has clotting properties, as shown in Figure 5-9.

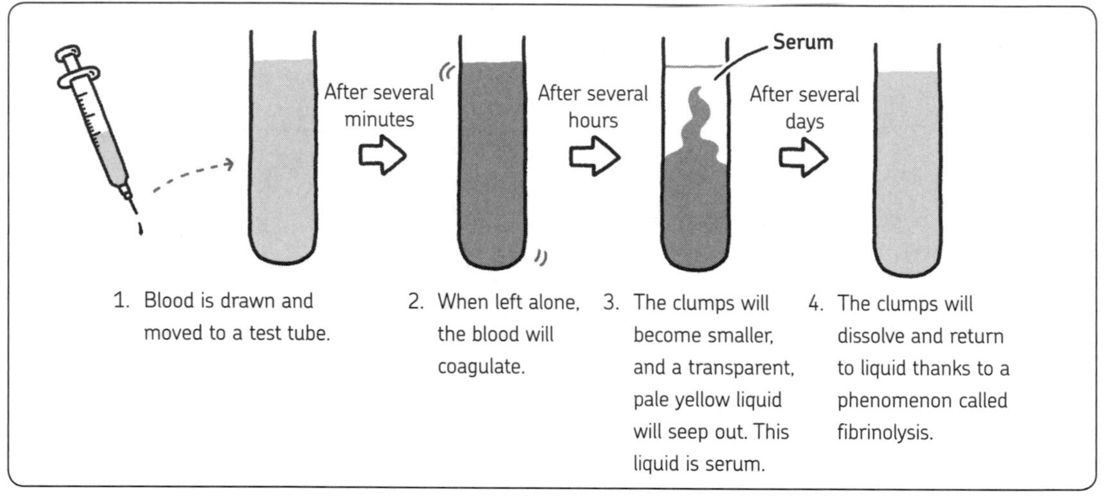

Figure 5-9: Coagulation in drawn blood

DID YOU KNOW?

After blood coagulates, a process called *fibrinolysis* will eventually break down and dissolve the blood clots. This mechanism prevents blood clots from spreading and causing problems in the body.

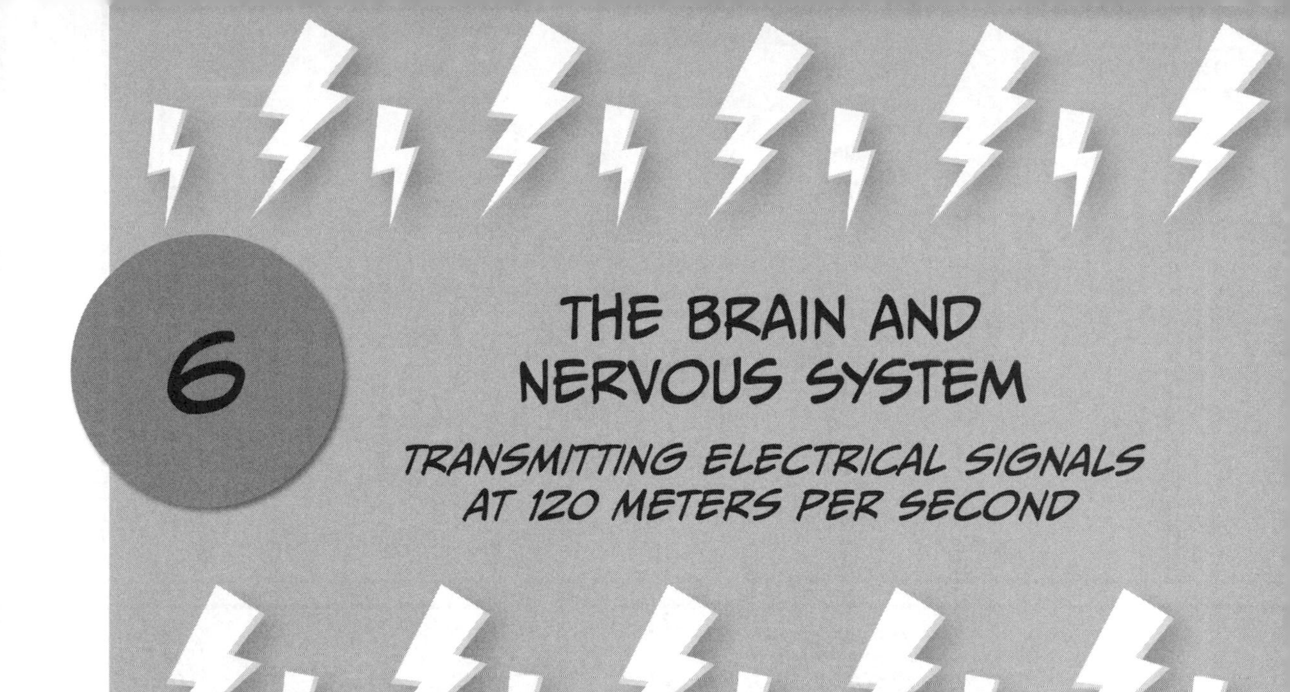

6 The Brain and Nervous System

Transmitting Electrical Signals at 120 Meters per Second

LOUNGE
OSAMU, DEAR BOY...
SQUEAK
UM, EXCUSE ME, DORM MOTHER...
PROFESSOR?

SORRY TO INTERRUPT...
...BUT WHEN DO YOU THINK THE AIR CONDITIONER IN MY ROOM WILL GET FIXED?
HUH? MS. KARADA?!

JUNIOR, DO YOU KNOW KUMIKO?
HI, PROFESSOR
YEAH, WE'VE MET A FEW TIMES...

UM, ABOUT THE AIR CONDITIONER...
I'M REALLY SORRY...
I'LL FIX IT TONIGHT!
HERE, COME ON IN AND HAVE SOME TEA.
IT'S MUCH COOLER IN THE LOUNGE.

TEA, HUH?
OK, I'M SORRY TO INCONVENIENCE YOU.

MS. KARADA, I DIDN'T KNOW YOU BOARDED HERE.
ARE YOU STILL CRAMMING FOR YOUR EXAM?
AH... SO COOL...

YES!
RIGHT NOW I'M STUDYING REALLY HARD ABOUT THE NERVOUS SYSTEM .
SO PROFESSOR, YOU'RE FRIENDS WITH MY DORM MOTHER?
AH, THANK YOU VERY MUCH!

NEURONS

I HEARD HER CALL YOU "JUNIOR."
STEAM
OUCH!!!!!
THROB THROB

YOU DIDN'T SCALD YOURSELF, DID YOU?
I SHOULD HAVE TOLD YOU IT WAS HOT TEA.
I LIKE A HOT DRINK WHEN IT'S HOT OUT.

* WHEN A NEURON IS STIMULATED, AN ELECTRICAL SIGNAL IS SENT DOWN THE AXON TO THE SYNAPSE AT THE END, WHICH THEN RELEASES CHEMICAL NEUROTRANSMITTERS TO STIMULATE THE NEXT NERVE. IF STIMULATION REACHES A CERTAIN THRESHOLD, THE NERVE ACTIVATES, TRANSMITTING THE SIGNAL. IF THE THRESHOLD ISN'T MET, THE NERVE REMAINS INACTIVE.

THE NERVOUS SYSTEM

Peripheral Nervous System

* AS WE LEARN, THE BRAIN ADAPTS BY MAKING NEW CONNECTIONS BETWEEN SYNAPSES, RATHER THAN BY CREATING NEW NEURONS. AS A RULE, NEURONS ARE HIGHLY SPECIALIZED AND DO NOT UNDERGO CELL DIVISION.

THE THREE KINDS OF PERIPHERAL NERVES OPERATE LIKE THIS.
Brain
Central nervous system
Spinal cord
Motor nerves
Sensory nerves
Relay point (autonomic ganglia)
Autonomic nerves
Skeletal muscles
Sense organs
Internal organs
THE MOTOR NERVES AND SENSORY NERVES TRANSMIT SIGNALS DIRECTLY, WHEREAS THE AUTONOMIC NERVES HAVE RELAY POINTS ALONG THE WAY.
THOSE RELAY POINTS ARE BUNDLES OF NERVE CELLS CALLED GANGLIA.
NOW, WHAT HAPPENED WHEN YOU TOUCHED THE HOT TEACUP AND IMMEDIATELY PULLED BACK YOUR HAND? THE SENSORY SIGNAL TOOK A SHORTCUT, TO MAKE YOU REACT QUICKER.
A SHORTCUT? HMM, LET ME THINK BACK...
I TOUCHED THE TEACUP AND WAS SURPRISED...
YEOW!
ACTUALLY, NO.
TUT-TUT
YOUR CONSCIOUS FEELING OF "SURPRISE" HAPPENED LATER.

FIRST, YOUR FINGERTIPS PERCEIVED AN EXTREMELY HIGH TEMPERATURE. THAT STIMULUS WAS TRANSMITTED THROUGH THE SENSORY NERVES AND TO THE SPINAL CORD.

1
Spinal cord

SO THAT'S WHEN I EXPERIENCED THE HEAT?
NOT YET...

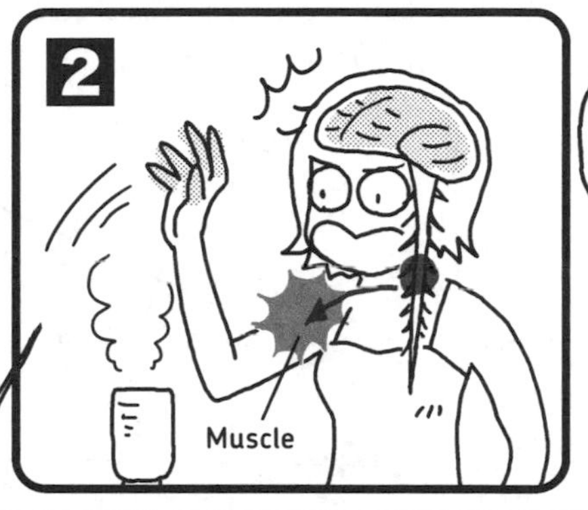
WHEN THE STIMULUS ENTERS THE SPINAL CORD, THE INFORMATION TAKES A SHORTCUT! INSTEAD OF WAITING FOR DIRECTIONS FROM THE BRAIN, THE SPINAL CORD TRANSMITS THE INSTRUCTION "CONTRACT!" STRAIGHT TO THE MOTOR NERVES, WHICH MAKES YOUR MUSCLES PULL YOUR HAND AWAY.
THIS IS CALLED A SPINAL REFLEX.
2
Muscle

SO AT THIS STAGE, THE HEAT SENSATION HAS NOT YET BEEN TRANSMITTED TO THE BRAIN?

RIGHT.
A SPINAL REFLEX OCCURS IN RESPONSE TO A STIMULUS THAT IS DANGEROUS TO THE HUMAN BODY, SUCH AS PAIN FROM A BURN.

THAT MAKES SENSE.
IT'S LIKE A DIRECT HOTLINE FOR A DANGER SIGNAL.

THEN...

3
Cerebrum

THE SENSORY INFORMATION ARRIVED AT YOUR CEREBRUM AROUND THE SAME TIME THAT YOU PULLED AWAY YOUR HAND.
THIS IS WHEN YOU CONSCIOUSLY RECOGNIZED THE HEAT AND PAIN.
I SEE.

HOW FAST DO THESE SIGNALS TRAVEL?
A STIMULUS IS TRANSMITTED ALONG A NERVE AT EXTREMELY HIGH SPEED.
ALTHOUGH IT VARIES DEPENDING ON THE STRUCTURE OF THE NERVE, THE TOP SPEED FOR FAST FIBERS IS AROUND 120 M/S.
THAT'S 432 KM/H!!
IT'S FASTER THAN THE *SHINKANSEN* BULLET TRAIN!
ふふふ
Bullet train (~300 km/h)
Motor nerve
IT'S AMAZING, ISN'T IT?
THE SIGNAL GOES FROM YOUR HAND TO YOUR BRAIN IN AN INSTANT!
YOU FELT STARTLED WHEN YOUR BRAIN RECOGNIZED THE PAINFUL HEAT SENSATION AND YOUR HEART RATE AND BLOOD PRESSURE PROBABLY ROSE.
MY AUTONOMIC NERVES FLEW INTO ACTION BECAUSE MY SYMPATHETIC NERVES WERE EXCITED!
THEY PERKED UP MY CIRCULATORY ORGANS.

YOUR CEREBRUM ISSUED A COMMAND TO LOOK AT THE HAND THAT RECEIVED THE HOT STIMULUS.

TO CARRY OUT THIS ACTION, YOU MOVED YOUR HAND, FACE, AND EYES.

YOUR RETINAS SENSED THE RED APPEARANCE OF YOUR HAND AND DELIVERED THIS INFORMATION THROUGH THE NERVES TO THE VISUAL CORTEX OF YOUR CEREBRUM. THEN YOUR CEREBRUM STARTED TO PUT THE INFORMATION TOGETHER: "MY HAND TURNED RED FROM THE HEAT OF THE TEACUP."

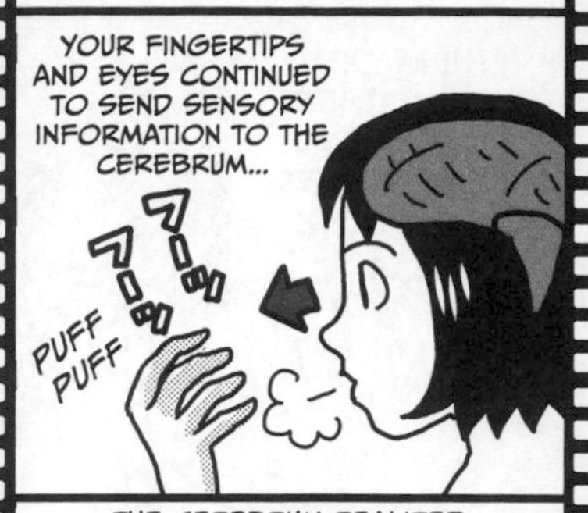

AND THIS BARELY TOUCHES ON THE COUNTLESS ACTIVITIES THAT ARE COORDINATED BY THE NERVOUS SYSTEM ALL THE TIME!

SO IF I RELATE THIS INFORMATION TO MY OWN BODY, I'LL UNDERSTAND AND REMEMBER IT BETTER?
THAT'S EXACTLY RIGHT!
DON'T WORRY, MS. KARADA. I THINK YOU'LL PASS YOUR MAKEUP EXAM WITH FLYING COLORS.
THANKS TO YOU, PROFESSOR...
KYAAA~
PHYSIOLOGY SEEMS SO INTERESTING NOW.
HAHA, IT'S BEEN AN HONOR.
BANG
RATTLE
BUT I THINK YOU'LL MAKE MORE PROGRESS IF YOU RELAX A LITTLE INSTEAD OF JUST STUDYING ALL THE TIME.
WHY DON'T WE GET OUTSIDE FOR A BIT?
HM
?
TA-DA!
WHA-?!
Summer Festival
IS HE ASKING ME OUT?!

EVEN MORE ABOUT THE NERVOUS SYSTEM!

You already know that the nervous system is divided into the central nervous system and peripheral nervous system. The nerves in the brain and spinal cord (the central nervous system) relay and collect information, make decisions, and issue instructions. Together, they can be thought of as the main control center of the body. Since the brain and spinal cord are such crucial organs, they are enveloped by meninges (membranes) and float in cerebrospinal fluid to protect them from impact. The brain collects and organizes all of our thoughts, emotions, perceptions, and behaviors, so let's begin with a discussion of the brain.

PARTS OF THE BRAIN

When you hear the word *brain*, you may think of the cerebrum—the main mass of the brain that is found under the skull—but that's not really all there is. The brain is made up of the cerebrum, diencephalon (interbrain), mesencephalon (midbrain), pons, medulla oblongata, and cerebellum, as shown in Figure 6-1.

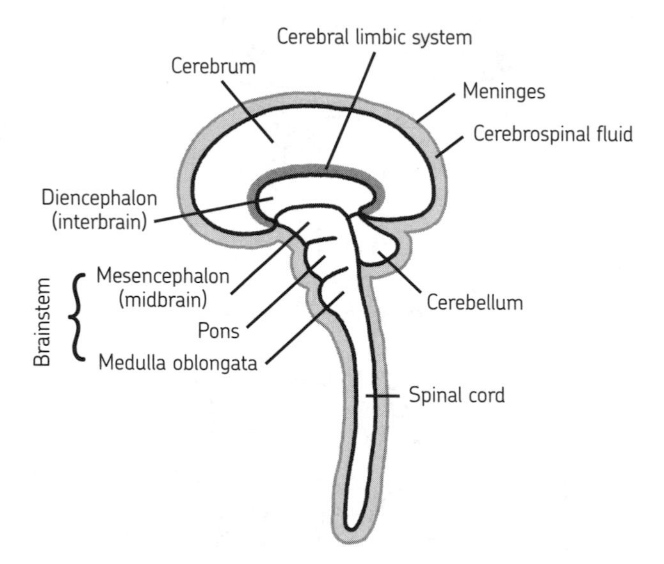

Figure 6-1: The central nervous system

The mesencephalon (midbrain), pons, and medulla oblongata are also called the *brainstem*. The brainstem is the innermost part of the brain and is responsible for the essential activities of life, such as respiration and circulation.

The part of the brain just above the brainstem, called the *cerebral limbic system*, is responsible for instinctive functions, such as appetite, sexual desire, pleasure, discomfort, and emotions.

The diencephalon (interbrain), between the cerebrum and the brainstem, includes the thalamus, hypothalamus, and pituitary gland. It functions as the control center of the autonomic nervous system and endocrine system.

STRUCTURE OF THE BRAIN

If you take a cross section of the brain, you'll see it divided into gray matter and white matter, as shown in Figure 6-2. The outer layer of gray matter is more formally called the *cerebral cortex*. It appears gray because of the neural cell bodies there, whereas the white inner layer is mostly made up of nerve fibers (axons). Those axons are white because they contain more fat tissue, which helps insulate the axons so that they can transmit signals more quickly.

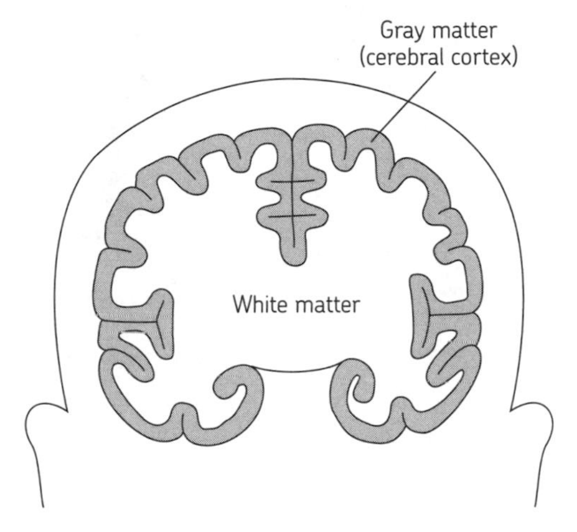

Figure 6-2: Cross section of the brain

The outermost layer of the cerebral cortex is called the *neocortex*. *Neo-* means "new," as the neocortex is the most evolutionarily recent addition to the brain. The neocortex accounts for about three-fourths of the human brain's mass.

The neocortex is responsible for the advanced cognitive functions that most distinguish human thinking from that of other animals. Worrying about failing an exam and studying for a retest are both jobs of the neocortex. Mastering the knowledge and skills needed to be a registered nurse, assessing patients and planning their care, and enjoying your time with friends are also all jobs of the neocortex.

Notice in Figure 6-2 how the surface of the cerebral cortex is folded into many wrinkles. This dramatically increases its surface area so that a lot more neocortex can fit inside the skull. Our high capacity for intelligence owes a lot to the wrinkles in our brain.

In addition to the neocortex, the cerebral cortex contains the *paleocortex* and the *archicortex*. The paleocortex and archicortex developed much earlier in the evolutionary timeline, and they are responsible for basic functions that we have in common with other animals, such as appetite, sexual desire, and the sensation of pain. Figure 6-3 shows how the brainstem, paleocortex/archicortex, and neocortex stack up across snakes, dogs, and humans.

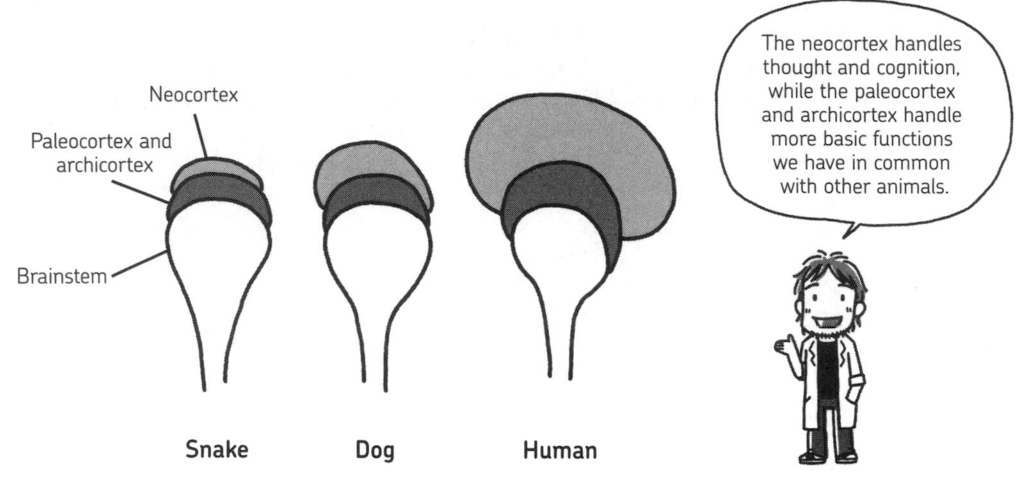

Figure 6-3: Differences between the cortexes of animals and humans

Getting back to the neocortex, different areas of this structure are responsible for different functions, such as vision, speaking, walking, running, other motor skills, and so on. This is called *localization* of brain functions, or *functional specialization*. Figure 6-4 shows the most important areas of functional specialization. The central sulcus marked in Figure 6-4 is a deep fold that runs across the middle of the brain, roughly from ear to ear. This feature separates the frontal and parietal lobes, and the motor and sensory cortexes.

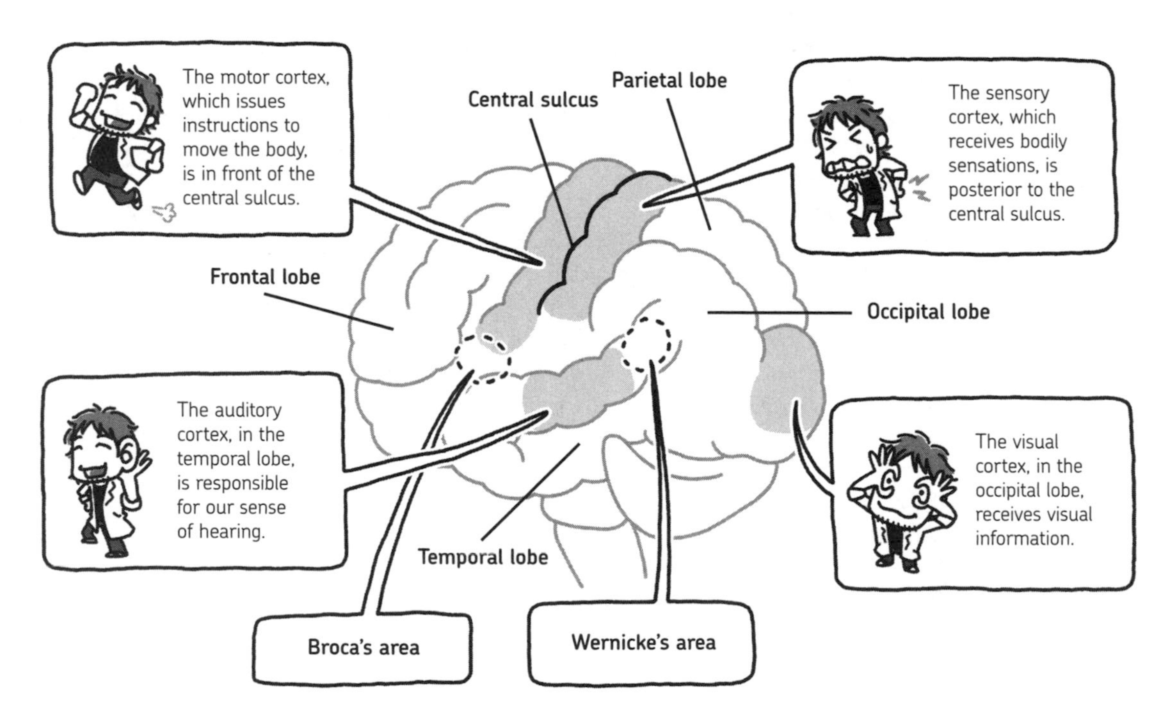

Figure 6-4: Functional specialization of the neocortex

BRAIN FUN FACTS

The left hemisphere of the cerebral cortex is responsible for movement and sensation on the right side of the body, and the right hemisphere is responsible for movement and sensation on the left side of the body.

Functions of the motor cortex and sensory cortex are further divided, with regions closer to the top of the brain responsible for the feet and regions closer to the temporal part of the brain (the side) responsible for the face and head.

So does this mean that even when I'm just having a conversation, several regions are performing different functions at the same time?

That's right. Let's try to think of which areas are required to handle language.

The methods we use to interpret language include reading characters and understanding sounds as words. Since each of these methods uses different sense organs and different types of information, separate locations are responsible for them.

The main area responsible for understanding language is *Wernicke's area*. However, when we speak we must shape the words by moving our mouth and tongue and jaw and so on. The area responsible for these functions is *Broca's area*. Both of these speech centers are in the left cerebral hemisphere. When brain damage causes problems with language, the symptoms differ depending on where the damage occurs. For example, if Broca's area is damaged, a person may not be able to speak coherently, or at all, even if the person still understands other people's speech.

The cerebrum issues instructions for telling the body to move. But isn't the cerebellum involved with movement as well?

Yes, the cerebellum coordinates your movements. It is below the cerebrum and behind the brainstem, compares the movement instructions issued by the cerebral cortex with the movements you actually make and issues signals to fine-tune the movement. As you repeatedly practice a movement, you become more skillful at it. This is the result of fine-tuning by the cerebellum.

BRAIN INJURIES

What about when the brain stops functioning? What's the difference between a vegetative state and brain death?

In a vegetative state certain parts of the brain are still functioning, whereas brain death is an irreversible state in which all brain functions are lost. This includes involuntary functions that keep the body alive, so when brain death occurs, the body dies too.

This is because not only is the person unable to speak or eat, but spontaneous respiration ceases and the heart quickly stops. When someone is in a vegetative state, on the other hand, the brainstem is still alive, and therefore respiration can occur and the heart continues beating. However, the person is not conscious and cannot respond. Figure 6-5 shows the difference in the brain between these two conditions.

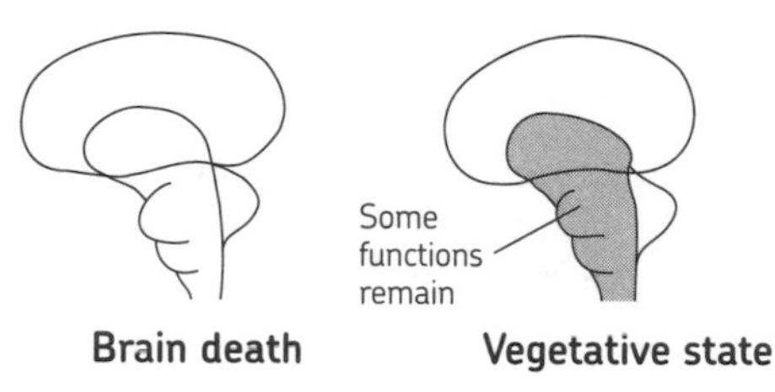

Figure 6-5: Difference between brain death and a vegetative state

The brain is very susceptible to oxygen deprivation. If respiration stops and the supply of oxygen is cut off, cells will begin to sustain damage in just 3 to 4 minutes. This is why it's so important to begin cardiopulmonary resuscitation as soon as possible if someone suffers cardiac arrest and the heart is unable to send oxygen-carrying blood to the brain.

BRAIN FUN FACTS

The brain consumes a large amount of oxygen because it constantly burns glucose as an energy source. In fact, glucose is normally the only energy source of the brain; although if glucose levels are too low, the brain may use molecules called ketone bodies as an alternative energy source.

THE SPINAL CORD

The main job of the spinal cord is to relay instructions issued from the brain to peripheral nerves and relay information from peripheral nerves to the brain. However, it has other important functions as well.

The *spinal cord* is an elliptically shaped cord approximately 1 centimeter thick extending from the bottom of the brain through the spinal column to the lumbar (lower back) region. When a baby first develops in the womb, the spinal column and brain start off as a single hollow tube. As this tube grows, the cells at the tip (head) increase and become the cerebral cortex (Figure 6-6), and the rest become the spinal cord.

Inside the spinal cord, nerve cells and nerve fibers form a bundle. Remember from earlier (see page 122), that neurons consist of nerve fibers (dendrites and axons) and cell bodies. Like the brain, the spinal cord is divided into white matter, which is mostly nerve fibers, and gray matter, which contains mostly cell bodies. But in the spinal cord, the pattern is reversed—the white matter forms the outer layer, and the gray matter is at the center of the cord.

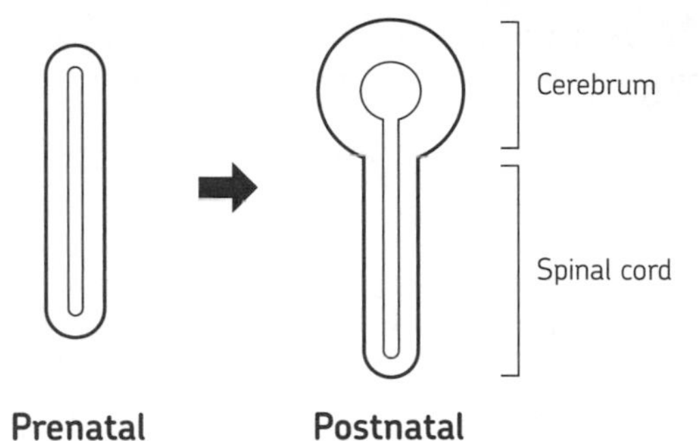

Figure 6-6: The cerebrum and spinal cord develop in the postnatal period.

In the spinal cord, the pathways for sending information and receiving information are separate so that sensory information and instructions from the brain do not interfere with each other, as shown in Figure 6-7.

Figure 6-7: How information is routed through the spinal cord

PATHWAYS THROUGH THE BODY

How exactly does the spinal cord relay instructions and information between the brain and peripheral nerves?

I'll explain this using Figure 6-8. Nerve fibers that descend along the spinal cord deliver the instructions from the brain to nerve cells in the gray matter in the front of the spinal cord (called the *ventral root* or *anterior root*). Spinal nerves extending from the anterior root deliver the instruction to the peripheral parts of the body. Meanwhile, nerve fibers that receive sensory information from the peripheral parts of the body enter the rear of the spinal cord (the *dorsal root* or *posterior root* of a spinal nerve) and deliver the information to the gray matter nerve cells there. Then those nerve fibers deliver the information to the brain. All of these signals are electrically transmitted along these various nerve fibers by way of action potential impulses. We can also call this process of propagating action potential *firing*.

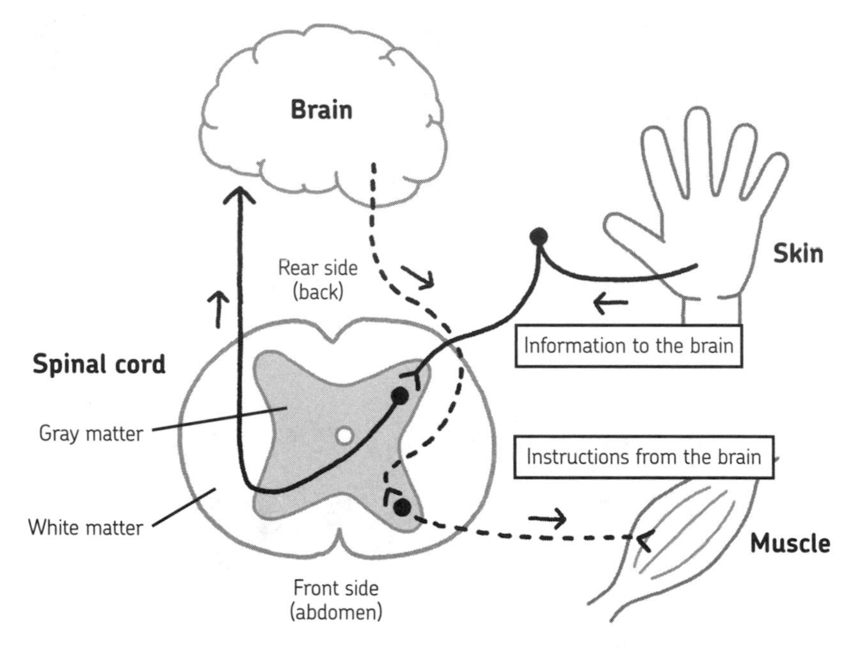

Figure 6-8: Nerve cell paths to and from the spinal cord

The nerves in the white matter of the spinal cord are separated according to their function. The *efferent* paths that transmit instructions from the brain and the *afferent* paths that transmit sensations to the brain are precisely divided. Those sets of fibers are called *conduction pathways*.

Most conduction pathways cross over between the left and right sides somewhere in the central nervous system. That's why the left hemisphere of the cerebral cortex is responsible for the right side of the body and the right hemisphere is responsible for the left side of the body.

So what kind of route does a signal take during a spinal reflex, like the one that caused me to pull my hand away the instant I touched the hot teacup?

Ah yes, in that case the signal takes a shortcut. An impulse indicating "hot!" was sent to your spinal cord. Since this is an *afferent* pathway, the impulse entered from the rear of the spinal cord. Normally, the impulse continues along an afferent pathway to another nerve cell, which passes the information to the brain. But instead, it took a shortcut through the spinal cord (Figure 6-9), and the information was delivered straight to the nerve cells of the *efferent* pathway on the front side. This caused your arm muscles to contract, pulling your hand away before you were even aware of what had happened.

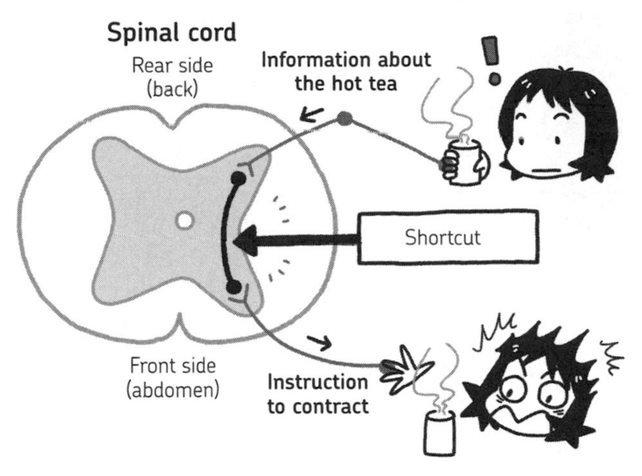

Figure 6-9: Spinal reflex shortcuts

DID YOU KNOW?

After birth, the spinal cord lengthens as the spinal column (vertebrae) develops. However, since bone growth outpaces spinal cord growth, in adults the area below the lumbar vertebrae contains only spinal fluid.

CRANIAL AND SPINAL NERVES

The brain and spinal cord form the central nervous system, and the nerves that link those central nerves with the peripheral parts of the body form the peripheral nervous system. Earlier, I told you that the peripheral nerves are classified into motor nerves that transmit movement instructions from the brain, sensory nerves that transmit sensory information to the brain from the peripheral parts of the body, and autonomic nerves that control internal organs. However, these nerves can also be divided along anatomical lines into cranial nerves stemming from the brain and spinal nerves stemming from the spinal cord (see Figure 6-10). The spinal nerves and brain nerves, along with the association nerves (which carry impulses between motor and sensory nerves), form the *somatic nervous system*.

There are 12 pairs of *cranial nerves*, each of which has a name and number. Most of our cranial nerves are either motor nerves that send movement instructions to the face, tongue, eyeballs, and so on or sensory nerves that transmit the five sensations from the head and skin. The *vagus nerve* is distinct from either of these, however. It branches down from the neck to regulate the internal organs of the chest and abdomen. The vagus nerve operates mainly as an autonomic nerve.

There are 31 pairs of *spinal nerves*, which stem from the spinal cord through openings between vertebrae. This group includes a mix of nerves that carry motor, sensory, and autonomic signals.

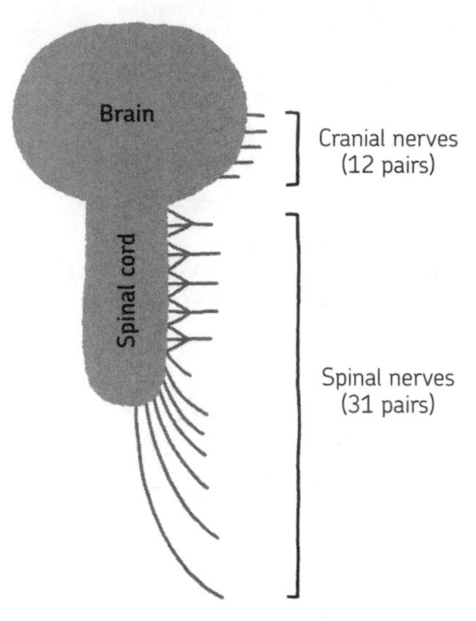

Figure 6-10: Cranial nerves and spinal nerves

Together, the cranial nerves and spinal nerves account for the movements and sensations of every inch in the body. Poke around anywhere on your body, from the tip of your little toe to the tip of your little finger to the top of your head, and you won't find a spot that has no sensation. That's pretty amazing, isn't it? Just think of how easily the handful of cords that connect a TV, DVD player, and cable box gets tangled up. Meanwhile, your nerves run throughout your entire body without ever getting any wires crossed, exchanging a huge volume of different kinds of information between the central and peripheral nervous system.

DID YOU KNOW?

When spinal nerves stem from the spinal cord, they divide into branches or merge with spinal nerves above and below them to create a meshlike structure called a *nerve plexus*.

THE AUTONOMIC NERVOUS SYSTEM

The word *autonomic* means involuntary and automatic. Many bodily functions are controlled by the autonomic nervous system without our being conscious of them. The autonomic nervous system has two major parts: the *sympathetic nervous system (SNS)*, which has to do with excitation and action, and the *parasympathetic nervous system (PNS)*, which has to do with calming and relaxation.

Different parts of the autonomic nervous system react depending on the situation. Let's suppose a herd of herbivores are eating plants on a savannah. They are relaxed and have no nearby predators to fear. In this relaxed state (nicknamed "rest and digest"), the parasympathetic nervous system is predominantly at work (see Figure 6-11). The neurotransmitter *acetylcholine* is released from parasympathetic nerves.

Figure 6-11: When you are relaxed, the parasympathetic nervous system is primarily in control.

Then some predators appear on the scene. The herbivores suddenly become nervous, and either run away or fight to protect themselves. In this agitated state (nicknamed "fight or flight"), the sympathetic nervous system is predominantly at work (see Figure 6-12). When a sympathetic nerve is excited, *norepinephrine* and *epinephrine* (adrenaline) are secreted from the adrenal medulla, maintaining the stimulated state of the body.

Figure 6-12: When you are scared or agitated, the sympathetic nervous system kicks in.

When the sympathetic nervous system becomes predominant, my heart starts thumping and my blood pressure rises.

That's right. The herbivore that believes it is about to be attacked by the predator must exert all its energy to run away or defend itself.

When this happens, the heart rate and blood pressure increase. The trachea expands so that lots of oxygen can be taken in, glycogen that was stored in the liver is broken down, and a large amount of glucose is released into the blood. At such times, you typically don't eat or excrete waste products. The blood flow to your digestive organs decreases, and the secretion of digestive fluids and movement of your alimentary canal slow down.

And while I'm unlikely to be attacked by a predator, I feel "attacked" by my test, right? I feel stressed!

Exactly. Although humans are not attacked by predators very often, they still respond similarly to threatening or unwelcome situations. Also, if this stress continues for a long time, a person's body and mind will become fatigued, as shown in Figure 6-13.

Figure 6-13: If the sympathetic nervous system is active for a long time, fatigue sets in.

It's essential for both the parasympathetic nervous system and the sympathetic nervous system to operate in a well-balanced manner (Figure 6-14). The autonomic nerves are distributed among the internal organs such as the heart and liver, internal secretory organs such as the pancreas and adrenal glands, the trachea and bronchial tube, the digestive system, the urinary bladder, and the arteries throughout the body. In most cases, the sympathetic nerves and parasympathetic nerves are distributed to exert opposite, complementary effects on all these systems.

Figure 6-14: Balance between the parasympathetic and sympathetic nerves

Is something wrong?

No . . . not at all!

The Enteric Nervous System

The third part of the autonomic nervous system, called the *enteric nervous system*, helps coordinate the actions of the gastrointestinal system. It is often called the "second brain" because it contains about 100 million neurons, more than the spinal cord or peripheral nervous system.

The enteric nervous system uses more than 30 neurotransmitters, and more than 90 percent of the body's serotonin is found in the gut.

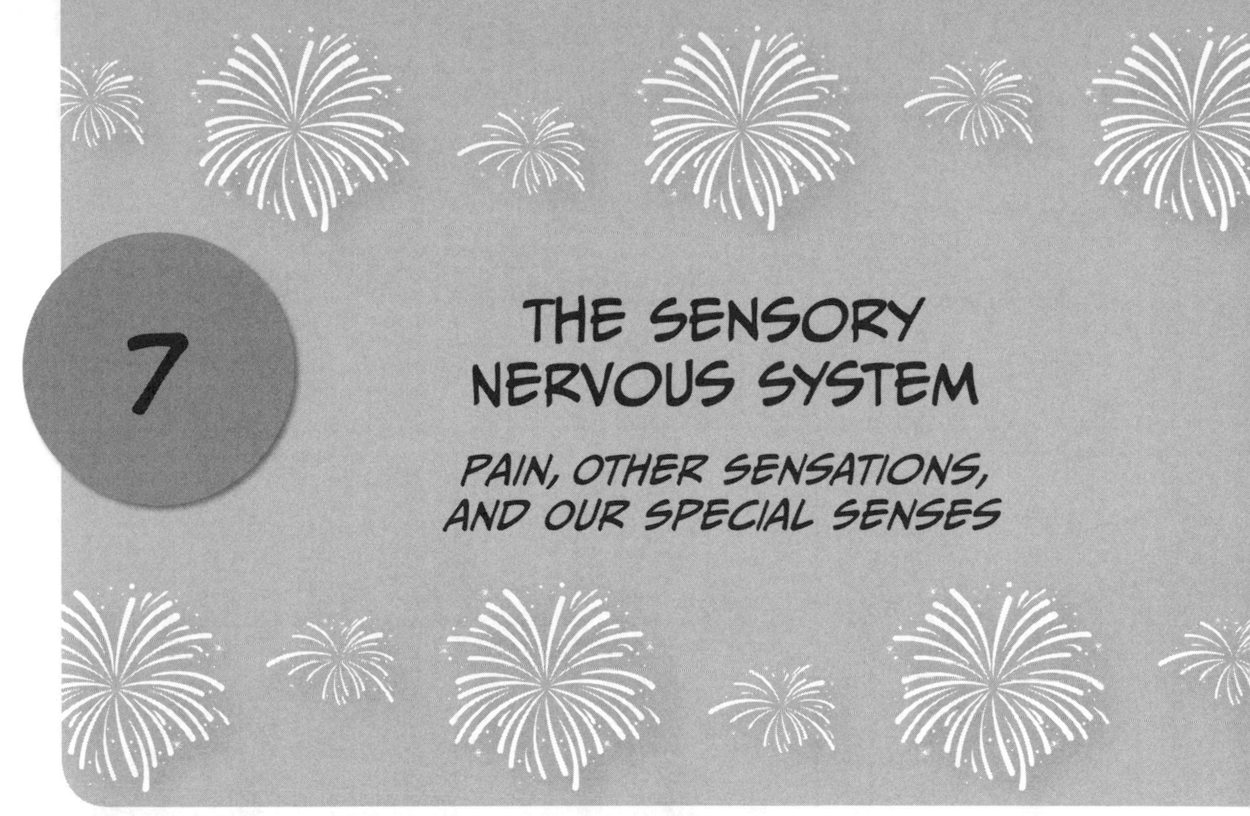

7 THE SENSORY NERVOUS SYSTEM

PAIN, OTHER SENSATIONS, AND OUR SPECIAL SENSES

YEAH, I LOVE THE PATTERN!
THIS DESIGN LOOKS A LITTLE LIKE A NEURON...
NEURONS?!

THAT'S RIGHT.
CLOSE YOUR EYES FOR A SECOND!
WHAT THE...?
えっ
HUH?
DON'T WORRY, DON'T WORRY...

JUST A SECOND...

THERE.
PLOP

WHAT IS THIS?

TODAY WE'RE GOING TO STUDY THE SENSORY ORGANS.
THE SKIN OF YOUR HAND IS PERCEIVING VARIOUS SENSATIONS NOW.
SO THE QUESTION IS...
WHAT SENSATIONS ARE THEY?
SENSATIONS?
UM, IT'S A LITTLE COLD AND...?
HEY!
YOU'RE TRYING TO *TEACH* ME, AREN'T YOU?!
DIDN'T YOU JUST SAY WE WERE TAKING A BREAK FROM STUDYING!!
WAH!
バタ
バタ
WHAT A SCARY FACE...
WHAT ELSE DO YOU FEEL?
...THE SURFACE IS A LITTLE ROUGH.
UMM...
OK, ROUGH... WHAT ELSE?
IT HAS SOME POINTY PARTS TOO.
GREAT, OK.
YOU CAN OPEN YOUR EYES NOW.
WHAT IN THE WORLD?!

* TURTLE SCOOPING AND GOLDFISH SCOOPING (*KINGYO-SUKUI*) ARE COMMON GAMES AT JAPANESE SUMMER FESTIVALS.

SEE HOW MUCH INFORMATION YOU LEARNED JUST BY HOLDING THAT LITTLE GUY? ALL THAT INFORMATION WAS PERCEIVED BY THE SUPERFICIAL SENSORY RECEPTORS OF THE SKIN.

SENSATIONS THAT ARE PERCEIVED BY THE SKIN ARE CALLED SUPERFICIAL SENSATIONS.

TYPES OF SENSATIONS

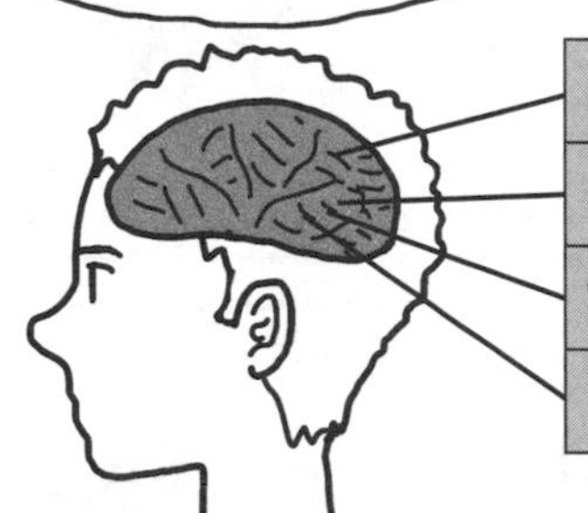

WHEN SOMETHING FEELS HEAVY OR LIGHT, THAT'S A DIFFERENT KIND OF SENSATION.
へえ

CLOSE YOUR EYES FOR A MOMENT.
えーっ
SO I'M GOING TO BE YOUR GUINEA PIG AGAIN?

WHAT'S HAPPENING NOW?
WELL...

IT SEEMS LIKE YOU'RE TRYING TO TAKE MY CANDY APPLE...
I'M NOT TAKING IT.
え?
BUT YOU CAN TELL THAT I RAISED YOUR HAND, RIGHT?
WELL, OF COURSE I CAN. THAT'S BECAUSE IT'S MY OWN HAND.

BUT WHAT'S GOING ON WHEN YOU PERCEIVE THE POSITION OF YOUR OWN ARM?

AH!
I GUESS THAT'S ITS OWN KIND OF SENSATION—HOW I PERCEIVE THE POSITION OF MY OWN BODY?
THAT'S RIGHT!

Proprioception (Position Sense)
I UNDERSTAND PERFECTLY WHAT'S HAPPENING WITH MY OWN HANDS AND FEET IN THIS POSE WITHOUT LOOKING AT WHERE THEY ARE.
PAT
PAT
PAT PAT
ぴた
ぴた
ぴたっ
THIS IS CALLED PROPRIOCEPTION (POSITION SENSE).
Kinesthesia (Sense of Body Movement)
AND IF I MOVE THIS WAY, I PERCEIVE THE KIND OF MOVEMENT MY BODY MADE.
THAT'S A PART OF PROPRIOCEPTION CALLED KINESTHESIA.
THESE SENSATIONS RELY ON RECEPTORS IN YOUR MUSCLES AND JOINTS, AS OPPOSED TO RECEPTORS IN YOUR SKIN.
AS A RESULT, THESE SENSATIONS ARE OFTEN CALLED *DEEP SENSATIONS*.
Position or Movement
Vibration
ANOTHER KIND OF DEEP SENSATION IS VIBRATION. YOUR MUSCLES USE THIS SENSATION TO PUSH BACK AGAINST BUMPS AND KEEP YOU UPRIGHT!
Deep Sensation
SO FAR WE'VE BEEN TALKING ABOUT TWO TYPES OF SENSATION.
Superficial sensation
Deep sensation
THAT'S RIGHT!
SUPERFICIAL SENSATIONS, WHICH WE MENTIONED EARLIER, AND DEEP SENSATIONS TOGETHER ARE KNOWN AS *SOMATIC SENSATIONS*.
I SEE.
Somatic sensation
Superficial sensation
Deep sensation
GOOD. I THINK YOU GET IT NOW.

HAVE SOME ICE CREAM AS A REWARD!
WOW
THANK YOU SO MUCH!
HERE I GO!
NOM NOM
WOLF WOLF
AAIIIEEEEEE!
YEESH... THAT LOOKS PAINFUL!
DO YOU KNOW WHERE YOU ARE FEELING THAT PAIN?
OW OW
IT'S NOT ON MY SKIN, SO IT'S NOT A SUPERFICIAL SENSATION...
OW OW
THAT'S RIGHT.
PAINS INSIDE THE BODY, LIKE HEADACHES AND STOMACHACHES, ARE CALLED *VISCERAL SENSATIONS.*
Somatic sensation
Superficial sensation
Deep sensation
Visceral sensation

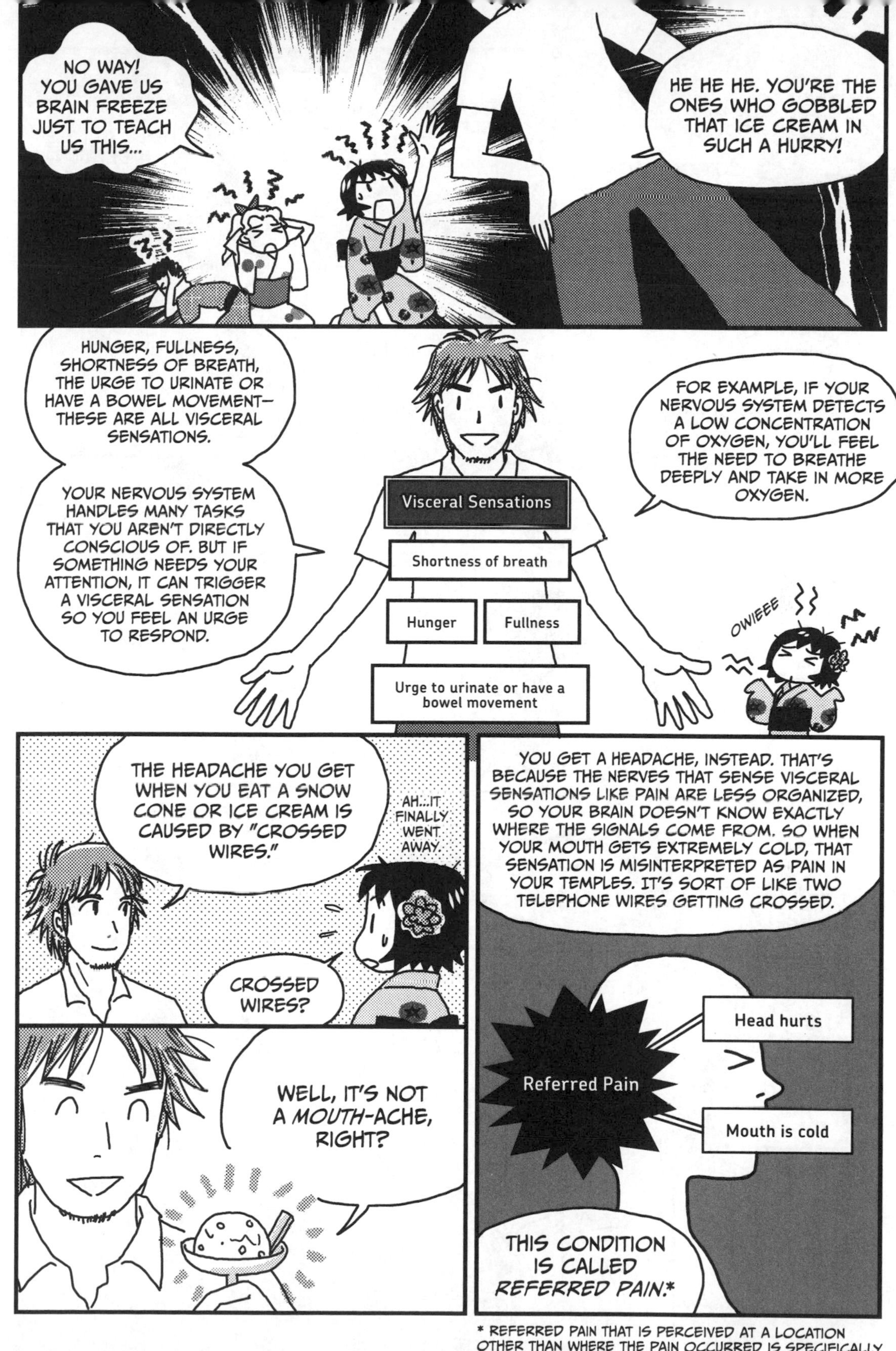

* REFERRED PAIN THAT IS PERCEIVED AT A LOCATION OTHER THAN WHERE THE PAIN OCCURRED IS SPECIFICALLY CALLED *RADIATING PAIN.*

THRESHOLDS AND SENSORY ADAPTION

OH...
THANK YOU.

NO PROBLEM.
よし
THERE

WERE YOU DRINKING SAKE?
YOUR FACE IS ALL RED.
「?」
HA HA NOOO...!
ははは
UM, UH...LET'S JUST GET BACK TO THE LESSON, PROFESSOR.

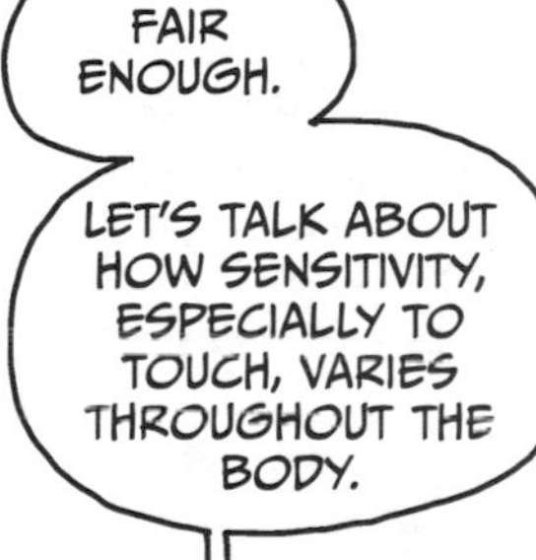
FAIR ENOUGH.
LET'S TALK ABOUT HOW SENSITIVITY, ESPECIALLY TO TOUCH, VARIES THROUGHOUT THE BODY.
IT ALL DEPENDS ON HOW MANY SENSORY RECEPTORS YOU HAVE IN A PARTICULAR AREA.

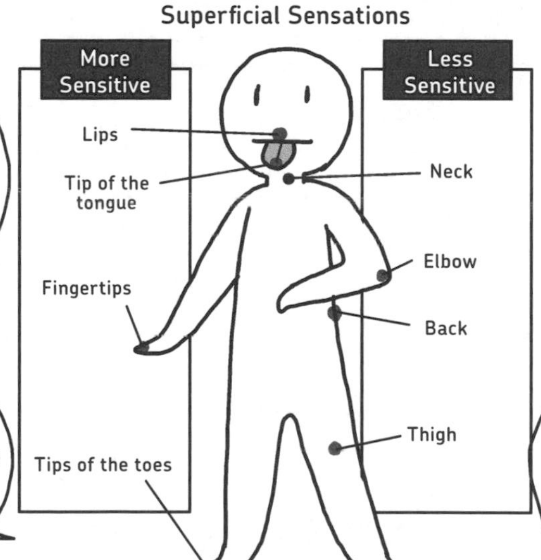
Sensitivity Differences for Superficial Sensations
More Sensitive
Lips
Tip of the tongue
Fingertips
Tips of the toes
Less Sensitive
Neck
Elbow
Back
Thigh

SOME ARE MORE DENSE IN CERTAIN AREAS, LIKE THE FINGERTIPS, LIPS, AND TIP OF THE TONGUE. THESE AREAS ARE MORE SENSITIVE AS A RESULT.
SENSORY RECEPTORS ARE MORE SPARSE IN THE BACK AND THIGHS, SO THESE AREAS ARE LESS SENSITIVE.

YEAH, THE SENSITIVITY OF MY FINGERTIPS IS CERTAINLY DIFFERENT THAN ON MY BACK. I GUESS THAT'S WHY I USE MY FINGERTIPS TO FEEL NEW THINGS.
THAT'S ALSO WHY BABIES USE THEIR MOUTHS TO FEEL NEW TOYS.

CERTAIN AREAS ARE MORE SENSITIVE THAN OTHERS, AND CERTAIN SENSATIONS HAVE A HIGHER PRIORITY, TOO—ESPECIALLY PAIN. YOU HAVE MORE RECEPTORS FOR TOUCH AND PAIN THAN YOU DO FOR GENTLE TEMPERATURES LIKE WARM AND COOL.

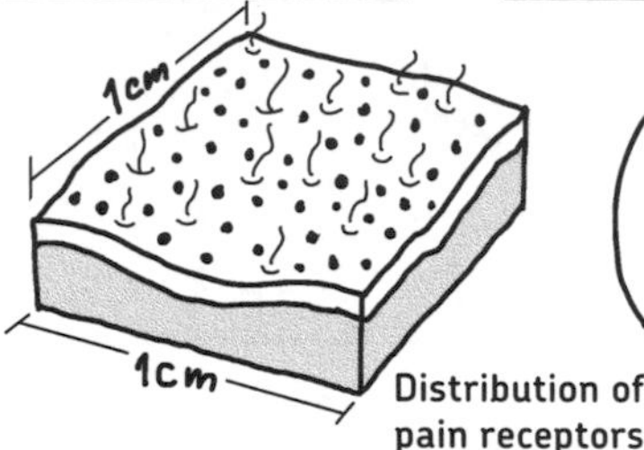

Distribution of pain receptors

THERE ARE MANY WAYS PAIN IS PRIORITIZED OVER OTHER SENSATIONS. FOR EXAMPLE, AN INJURY CAN CAUSE THE RELEASE OF CHEMICALS TO MAKE THE SURROUNDING AREA MORE SENSITIVE TO PAIN. THIS IS CALLED *HYPERALGESIA*.

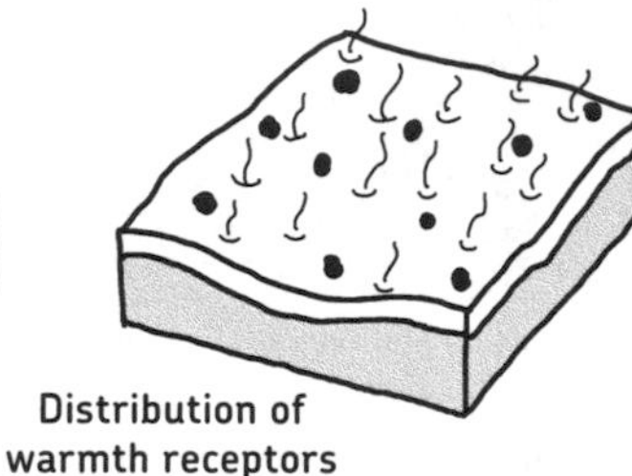

Distribution of warmth receptors

IN OTHER WORDS, HYPERALGESIA LOWERS THE PAIN THRESHOLD. THE MINIMUM LEVEL OF STIMULUS THAT CAUSES A SENSATION LIKE PAIN OR COLD TO BE PERCEIVED IS CALLED THE *THRESHOLD VALUE*.

Threshold Value		
	Low	Felt even with a weak stimulus (sensitive)
	High	Felt only with a strong stimulus (insensitive)

A LOW THRESHOLD VALUE MEANS THAT THE SENSATION IS FELT EVEN IF THE STIMULUS IS WEAK.

AND A HIGH THRESHOLD VALUE MEANS THAT THE SENSATION CANNOT BE FELT UNLESS THE STIMULUS IS STRONG.

IMAGINE WE APPLY A STIMULUS, LIKE POKING YOU, VERY GENTLY AT FIRST. THEN WE POKE A LITTLE BIT HARDER EACH TIME.

Strong

Strength of Stimulus

Weak

AT FIRST, THE STIMULUS IS SO SMALL THAT IT DOESN'T REGISTER AS A SENSATION.

WHEN YOU FIRST SENSE THE POKE, THAT IS THE THRESHOLD VALUE FOR YOUR TOUCH RECEPTORS. WHEN IT STARTS TO HURT IS THE THRESHOLD VALUE FOR PAIN RECEPTORS.

I SEE.

YOUR SENSITIVITY TO A STIMULUS CAN CHANGE OVER TIME, TOO.
FOR EXAMPLE, IF THERE IS A STIMULUS ABOVE THE THRESHOLD THAT REMAINS CONSTANT—LIKE IF I POKE YOU AND KEEP MY FINGER THERE—AFTER A WHILE YOU MAY NO LONGER SENSE IT.
Doesn't hurt anymore
Sensory Adaptation
Threshold value
Still doesn't hurt
THIS IS AN EXAMPLE OF SENSORY ADAPTATION.
MS. KARADA?
ん?
クルッ
TURN
HEY, PROFESSOR! COME OVER, QUICK!
ドーン
BOOM
HUH? WHERE'D THOSE TWO GUYS GO?
THEY MUST HAVE GOTTEN LOST...
OOOO, WHAT A SPECTACULAR SHOW...
MS. KARADA, IT'S YOUR FIRST TIME AT THE FESTIVAL, ISN'T IT?
ドーン
BOOM
パラパラ
POP POP
YES, IT'S BEAUTIFUL.
I HOPE IT WAS A NICE STUDY BREAK...EVEN IF WE STUDIED A LITTLE...
WELL, I STILL HAD A GREAT TIME TONIGHT.

EVEN MORE ABOUT THE SENSORY NERVOUS SYSTEM!

So far we've discussed superficial, deep, and visceral sensations. But there is a fourth major category of sensations: special senses, which include the senses of sight, hearing, balance, smell, and taste. All of them have specialized organs associated with them, such as the eyes, ears, and nose. Let's go through each of these special senses.

SIGHT AND THE EYE

Let's start with sight. The basic structure of the eye is a lot like that of a film camera (see Figure 7-1). The *crystalline lens* of the eye is like a camera lens, the *iris* the aperture, and the *retina* the film.

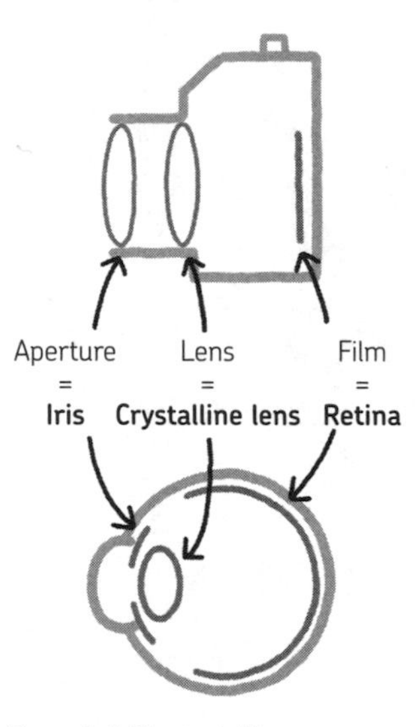

Figure 7-1: The eye is like a camera.

Figure 7-2 shows the structure of the eye in more detail. Light that enters the eye is refracted (or bent) by the *cornea* and lens, and is projected onto the retina as an image, which is both upside-down and backward. Two types of cells, called cones and rods, are tightly arranged in the retina. These cells can perceive when light hits them and send signals to the cerebral cortex through the *optic nerve*. Your brain then interprets the signals as light and puts together the image, reversing it to the correct orientation.

Is there a particular reason why humans, along with so many other animals, have two eyes?

Have you ever had to wear an eye patch? You might have noticed that you lost your *depth perception*, or sense of distance. When you focus on an object with both eyes, the images perceived by each eye are slightly offset from each other horizontally. How much these images are offset depends on how close or far away the object is. Your brain also analyzes a few other clues (like size, texture, and so on) to determine depth information so that you can see objects three-dimensionally.

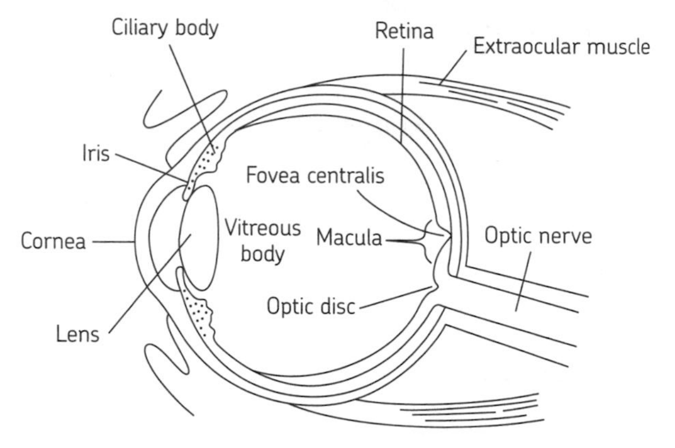

Figure 7-2: Structure of the eye

Light information is delivered to the visual cortex (which is part of the cerebral cortex), where the two images from each eye are integrated so that the object is "seen." But the light gets from the retina to the visual cortex by slightly different routes, depending on which eye and which part of the eye, received the information. The image projected onto the outer half of the retina (closer to your ears) is sent to the visual cortex on the same side of the brain as the eye that perceived it. Meanwhile, the image projected onto the inner half of the retina (closer to your nose) is sent across to the visual cortex on the opposite side of the brain. The part of the brain where these routes cross over is called the *optic chiasm* (Figure 7-3).

This means that if you hold up both hands in front of you, the image of your right hand is sent to your brain's left hemisphere (which controls that hand!), and the image of your left hand is sent to the right hemisphere (which controls that one!). In other words, when you catch a ball coming from your left with your left hand, it's your right brain that both sees the ball and directs your left hand to catch it. This way, you put both sides of your visual field together as one picture, while the parts of your brain that need to see and respond to objects in front of you are close together.

DID YOU KNOW?

When you look at something, you don't just turn your face; your eyeballs move, too. The rotation of your eyeballs is controlled by a total of six muscles, called the extraocular muscles.

The pupil is the hole in the iris that lets in light. It contracts when there is lots of light, and it becomes larger in the dark. The pupils are controlled by autonomic nerves, and in the absence of disease they are the same size in both eyes.

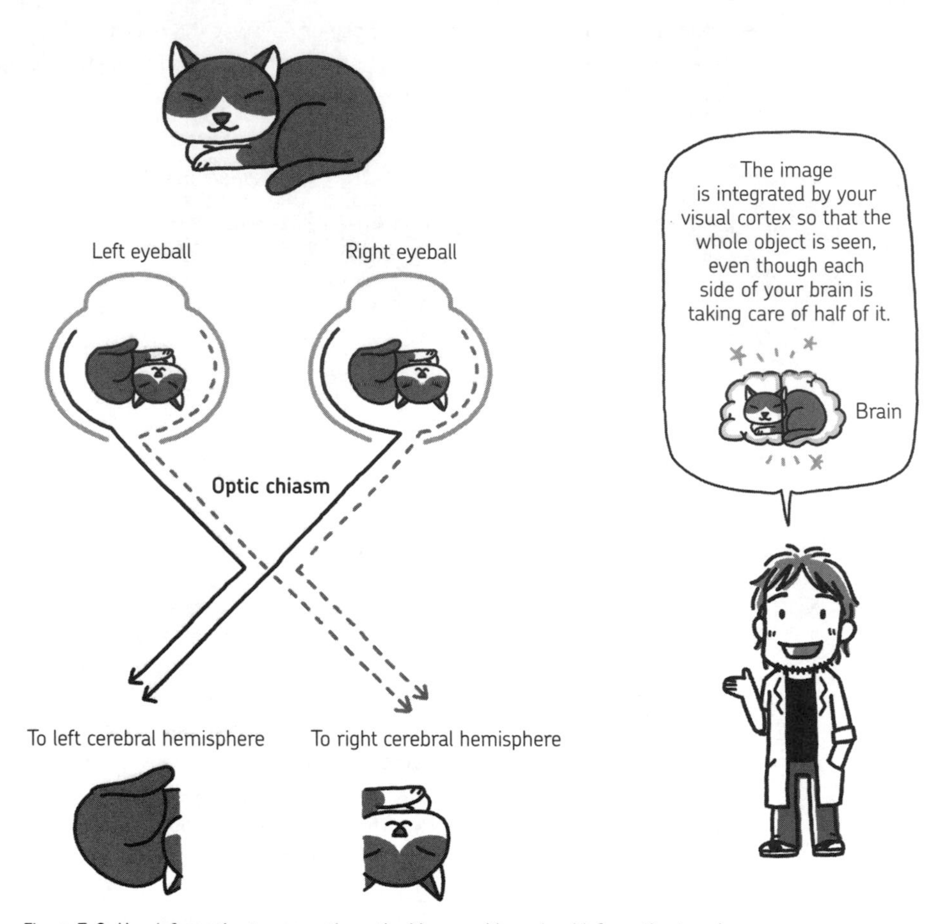

Figure 7-3: How information traverses the optic chiasm and how visual information travels

MYOPIA AND OTHER COMMON CONDITIONS

But the eye doesn't always work perfectly. Myopia occurs when the lens can't adjust its thickness to focus properly, right?

Yes, that can be one cause of myopia. *Myopia* (nearsightedness) is when you cannot focus on distant objects—its literal meaning is "trying to see like a mole." *Hyperopia* (farsightedness) is when a clear image cannot be projected on the retina, so that you cannot focus on objects closeby.

Both of these conditions often arise because the size of the eyeball has changed. If the eyeball gets longer, the distance from the lens to the retina increases, and it becomes more difficult to see distant objects (causing myopia). On the other hand, if the eyeball is too short, the distance between the retina and the lens decreases, and nearby objects are difficult to see (causing hyperopia).

While we're talking of common eye troubles, *astigmatism* is a condition in which the refractive index of the cornea differs in the vertical and horizontal directions; thus, the lens bends light as an oval, rather than as an even circle, distorting the image.

Finally, *presbyopia* (so-called age-related farsightedness) is a condition in which the thickness of the crystalline lens can't adjust by stretching and relaxing any more. This is often because of aging, but this can also happen due to other factors.

COLORS AND LIGHT IN THE EYE

How do we distinguish colors?

That's the role of rods and cones in the retina. *Cone cells* distinguish colors. There are three types of cone cells, and each one corresponds to one of the primary colors of light: red, green, or blue. *Rod cells*, on the other hand, only distinguish light from darkness, regardless of the color of the light. The rods are more sensitive, however, and signal one another if very little light is entering the eye. As a result, in darker situations, the cones cannot detect any light, and we see mostly using the rods, giving the sensation that things are less colorful or more "black-and-white." The rods and cones are arranged throughout the retina, but they are particularly dense in and around the *macula*, the central portion of the retina where the lens projects the middle of an image. The center of the macula, the fovea centralis, is where the center of the field of vision is focused.

Visual Acuity

Visual acuity describes the sharpness and accuracy of a person's vision. It measures the ability to recognize light and darkness, color, distance, and moving objects.

Visual acuity is often tested using an eye chart of letters and symbols arranged from largest to smallest, which tests whether an eye can distinguish details at a certain distance. Other tests can distinguish the ability to recognize color, distance, and moving objects, as well as the presence of any problematic "blind spots" in the visual field, or regions in which a person's eyes or brain cannot detect light.

HEARING AND THE EAR

Your ears are responsible not only for your sense of hearing but also for your sense of balance. Let's take a look at the ear's structure (Figure 7-4).

The ear is divided into the outer ear, middle ear, and inner ear. The outer ear consists of the auricle (the main bulk of cartilage attached to the side of your head) and the ear canal. The eardrum (tympanic membrane) and the auditory ossicles behind it make up the middle ear. The auditory ossicles are a small set of bones called the malleus, incus, and stapes (Latin for hammer, anvil, and stirrup, respectively). The inner ear is embedded deep inside the skull, and it includes the semicircular canals and vestibule, which make up the vestibular system, and the snail-shaped cochlea. Although all of these parts help perceive sound, only your inner ear (specifically the vestibular system) is involved in your body's perception of balance.

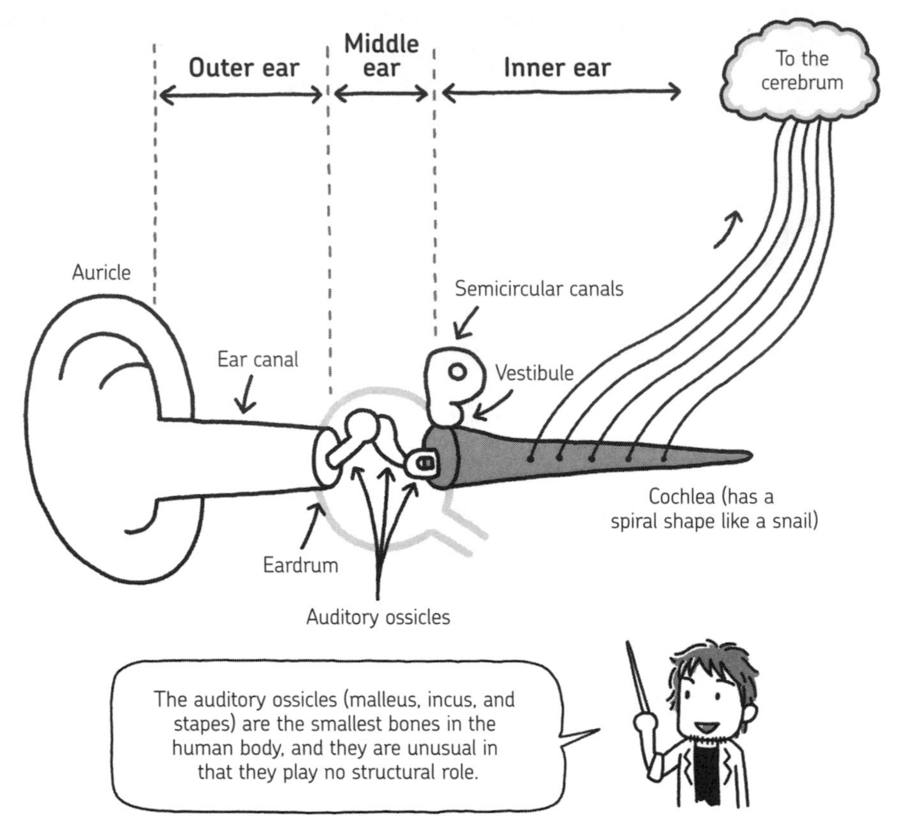

Figure 7-4: Structure of the ear

Sound is just vibrations of the air. The ear is an apparatus that amplifies those vibrations and then converts them into nerve impulses. The shape of the auricle helps reflect sound into the ear, where the tubular ear canal helps direct those vibrations toward the eardrum, which vibrates in response. The three tiny auditory ossicles inside the middle ear make those vibrations into larger movements and transmit them to the inner ear. The cochlea in the inner ear is filled with lymph fluid, and when that fluid vibrates, receptor cells inside the cochlea are able to distinguish frequencies and convert them to nerve impulses.

What causes hearing loss?

There are several kinds of hearing loss depending on where the problem occurs in the ear.

Since the outer ear and middle ear conduct sound, they are sometimes referred to as the sound conduction system. A problem in these parts of the ear is called *conductive hearing loss*. Specific examples include blockage of the outer ear, perforation of the eardrum, or loss of mobility in the auditory ossicles.

Conductive hearing loss can be alleviated by products that aid bone conduction. Bone conduction is the delivery of vibrations to the inner ear through the skull (rather than through the outer and middle ear), as shown in Figure 7-5. Therefore, as long as the inner ear is still intact, sound can still be perceived. Earphones that use this technique have recently appeared on the market.

Since the inner ear "perceives sound," it is sometimes referred to as the sound perception system. The inability to recognize sound due to a problem in the nerves in the inner ear or cerebrum is called *sensorineural hearing loss.*

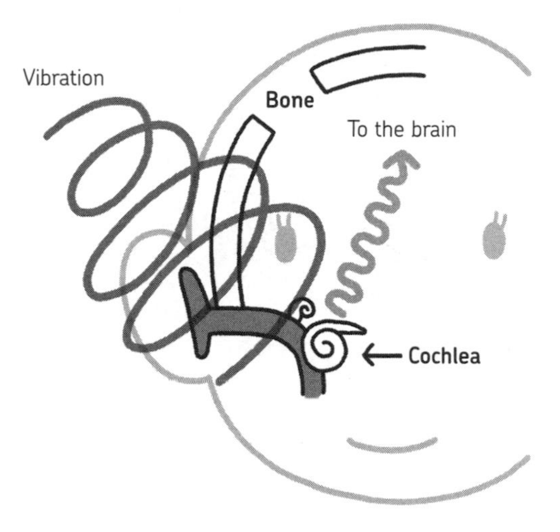

Figure 7-5: Bone conduction

BALANCE AND THE INNER EAR

Tell me more about the vestibular system. How does it perceive balance?

The inner ear perceives two types of motion: rotational motion and inclination of the head. Rotational motion is perceived by the *semicircular canal system*, while the inclination of the head is perceived by the *vestibule*, which is the part below the semicircular canals.

Ears Help Equalize Pressure!

For the eardrum to move freely, the pressure on the outer ear (atmospheric pressure) and the pressure on the middle ear needs to be similar. The Eustachian tube, which connects the middle ear to the back of your nasal cavity, lets fluid drain or air move back and forth to equalize the pressures. But if the air pressure around you changes rapidly, as during an airplane ride or while scuba diving, the difference in pressures may cause a ringing sound. Some people find that making an effort to blow their nose while holding their nostrils shut can push air into the Eustachian tube, popping it open and equalizing the pressures to stop the ringing.

The semicircular canals are three loops arranged at perpendicular angles to each other. When your head turns in any direction, lymph fluid inside the semicircular canals pushes and bends specialized nerve cells at the base of each loop. These nerves, in turn, send signals to the brain that cause a feeling of motion in that particular direction. This is shown in Figure 7-6.

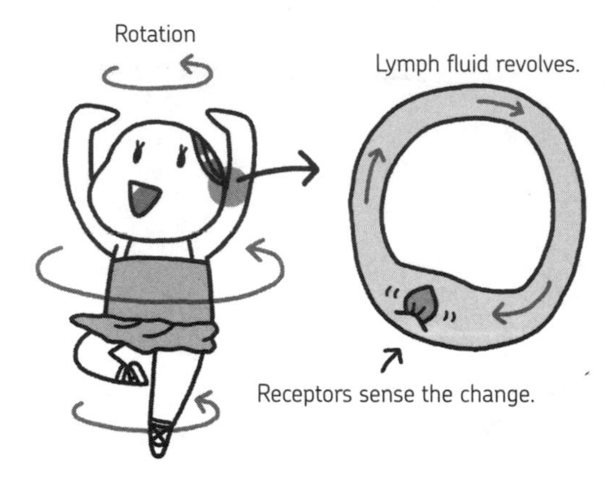

Figure 7-6: As the lymph fluid flows, specialized nerve cells detect the rotation.

Meanwhile, the vestibule contains an apparatus for detecting which way is up (see Figure 7-7). In the vestibule, tiny particles ride on a similar fluid so that when you tilt your head by bending your neck, the fluid is moved by gravity, and nerve cells perceive the direction that gravity is pulling "down." With this information, your brain can tell which way is "up" and, by comparison, the tilt of your head.

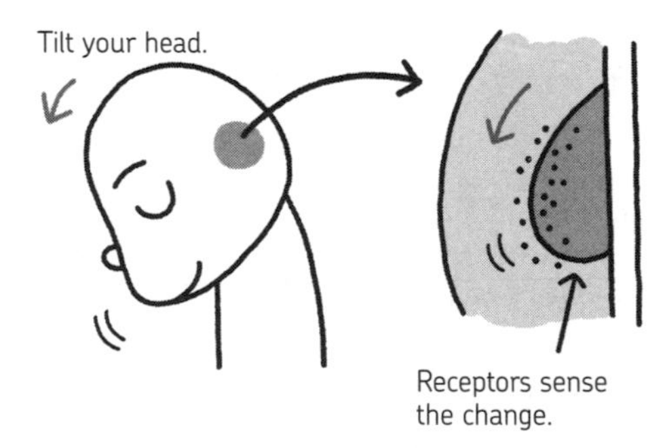

Figure 7-7: Particles ride on the fluid and detect how the direction of the force of gravity changes as your head tilts.

SMELL AND THE NOSE

Smell is perceived by the olfactory epithelium, an area about the size of a fingertip located at the top of the nasal cavity (see Figure 7-8). This area is packed with chemical receptors.

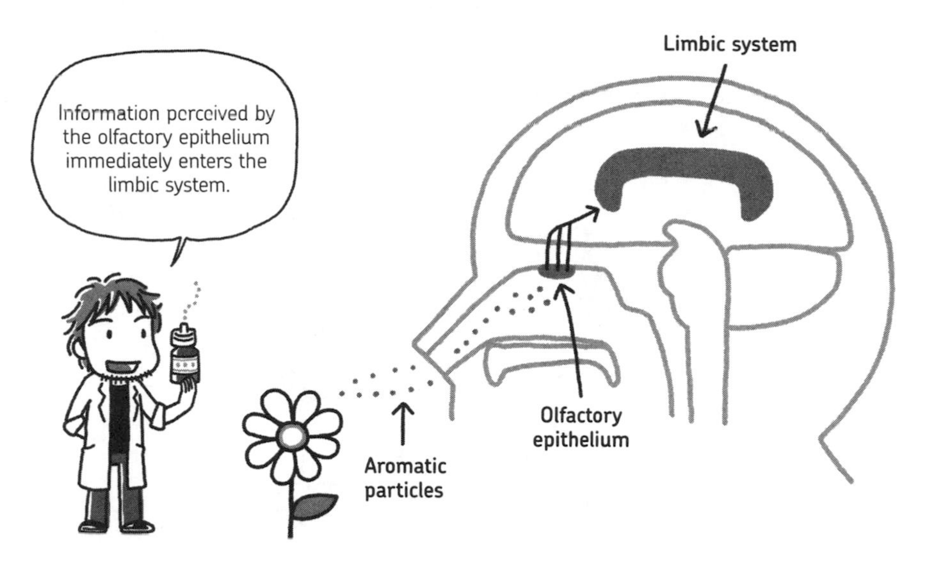

Figure 7-8: The mechanism for the sense of smell

When an *odorant* (any chemical compound that these receptors can detect) enters the nasal cavity and comes in contact with the olfactory epithelium, these cells send sensory information to the brain.

This sensory information passes through the roof of the nasal cavity and enters the limbic system (see "Parts of the Brain" on page 130), where it is processed by the olfactory bulbs in the brain. The limbic system also contains structures that are responsible for basic drives and emotions such as appetite and sexual desire, motivation, and pleasure or displeasure. Because the sense of smell is processed so close to this emotional center of the brain, certain smells often have strong ties with specific emotions.

Smells can also be closely linked to memories. Has the scent of a certain familiar fragrance ever instantly brought to mind a scene from your past in vivid detail? Once again, this is related to the physical proximity of the olfactory bulbs and other structures in the limbic system that deal with memory (see Figure 7-9).

Figure 7-9: Aromas can evoke strong memories.

When your nose is clogged because of a cold or hay fever, you can't taste your food as well. That's because a major part of the sensation we think of as "taste" is really smell. Although our sense of taste is obviously perceived by the tongue, we cannot recognize the full "tastiness" of food without our sense of smell, as shown in Figure 7-10.

Figure 7-10: Your sense of taste is dulled when you can't smell.

Although our sense of smell is no match for that of a dog, it's still quite impressive. Humans have the ability to distinguish more than 10,000 distinct odors.

Sensory adaptation readily occurs for our sense of smell. That is, if you are surrounded by an odor, you may quickly grow accustomed to it and become temporarily unable to perceive it. However, even after adapting to a smell, you might still be quite sensitive to changes in the intensity of the odor and remain perceptive of other odors.

TASTE AND THE TONGUE

Humans can perceive five tastes: saltiness, sweetness, bitterness, sourness, and umami (a "savory" taste, such as of the amino acids glutamate and aspartate). It used to be thought that each taste was only perceived by a certain area of the tongue (for example, saltiness was perceived only by the tip of the tongue and bitterness by the back part), but this theory has since been repudiated. Your sense of taste is important because it gives you a chance to perceive both the nutrients and toxins or poisons in food.

The sense of taste is mainly perceived by taste buds on the surface of the tongue. Each taste bud is like a little pocket that contains cells that perceive flavors. When flavor components such as salt or sugar mix with saliva and spread throughout the mouth, they are perceived by taste buds, which send this sensory information to the brain.

Since a taste bud is a tiny apparatus (see Figure 7-11), large molecules of food, like carbohydrates in rice or bread, need to be broken down into smaller molecules in order to be tasted. Chewing and enzymes like amylase in your saliva help break down food for your taste buds.

Figure 7-11: Structure of a taste bud

Where are taste buds located on the tongue?

Taste buds are particularly numerous in the little bumps called papillae on the surface of the tongue. However, taste buds are also found in the mucus membrane of the mouth and throat.

MOUTH FUN FACTS

There are four kinds of papillae on the tongue. Filiform papillae are the most numerous kind, and they are largely responsible for the roughness of the tongue's surface. However, unlike the other kinds of papillae, they do not contain any taste buds. They are just there to grip and break up food!

The temperature of food affects how the taste buds perceive flavors. The warmer a food is when served, the more sensitive you might be to its sweetness or bitterness. This is why melted ice cream can taste too sweet and warm beer might taste extra bitter.

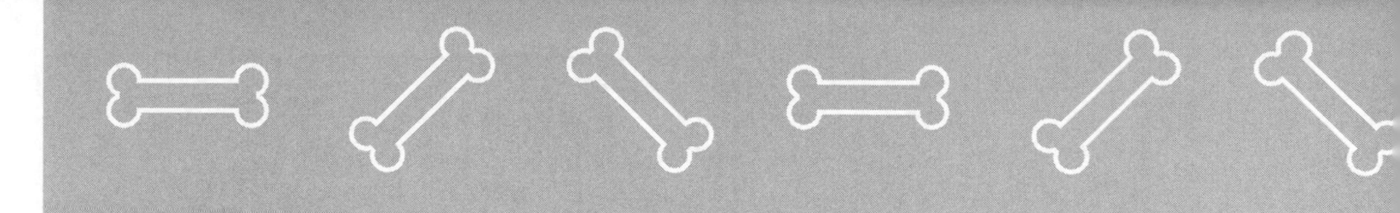

8 THE MUSCULOSKELETAL SYSTEM

MUSCLES, BONES, AND JOINTS

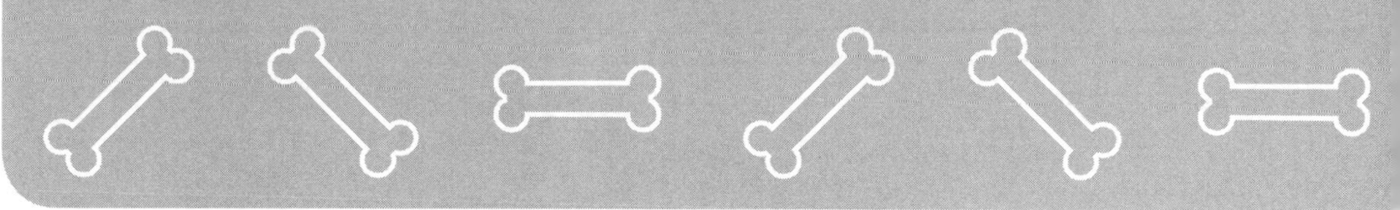

YOU BET!
YOU MUST BE TRAINING HARD.
YOU'RE DOING SOME STRETCHES, I SEE.
YEAH...
I WANT TO INCREASE MY MUSCLE ELASTICITY SO I DON'T INJURE MYSELF!
GOOD FOR YOU!
I BET YOU KNOW ALL ABOUT MUSCLE FIBERS AND HOW THEY USE ENERGY, RIGHT?
WELL, UM... PRETTY MUCH.
JUST IN CASE, CAN YOU GO OVER IT?
I'M A LITTLE RUSTY.
MUSCLE FIBERS
OF COURSE!
DO YOU KNOW WHAT'S HAPPENING WHEN MUSCLES CONTRACT?
HUH?
WELL, WHEN YOU FLEX REALLY HARD, THE MUSCLE GETS ROCK SOLID.

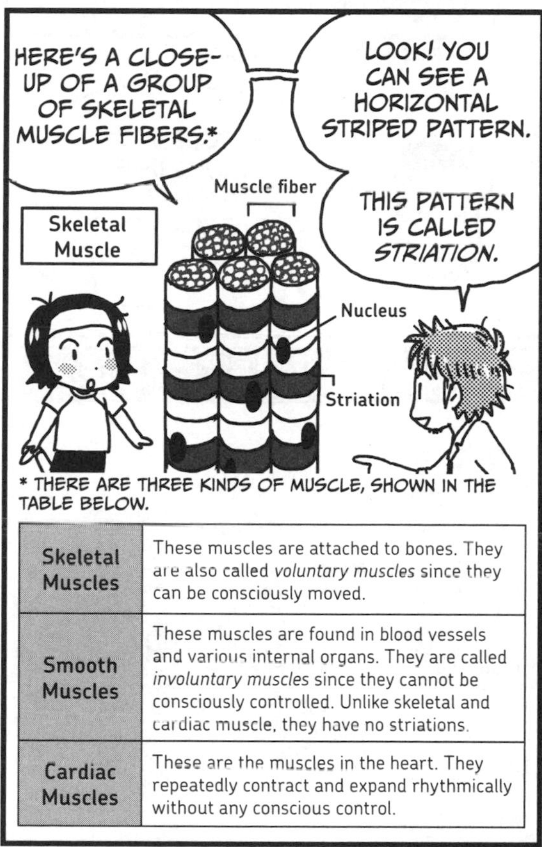

* THERE ARE THREE KINDS OF MUSCLE, SHOWN IN THE TABLE BELOW.

Skeletal Muscles	These muscles are attached to bones. They are also called *voluntary muscles* since they can be consciously moved.
Smooth Muscles	These muscles are found in blood vessels and various internal organs. They are called *involuntary muscles* since they cannot be consciously controlled. Unlike skeletal and cardiac muscle, they have no striations.
Cardiac Muscles	These are the muscles in the heart. They repeatedly contract and expand rhythmically without any conscious control.

* SEE "ATP AND THE CITRIC ACID CYCLE" ON PAGE 74.

Muscles Are Powered by ATP

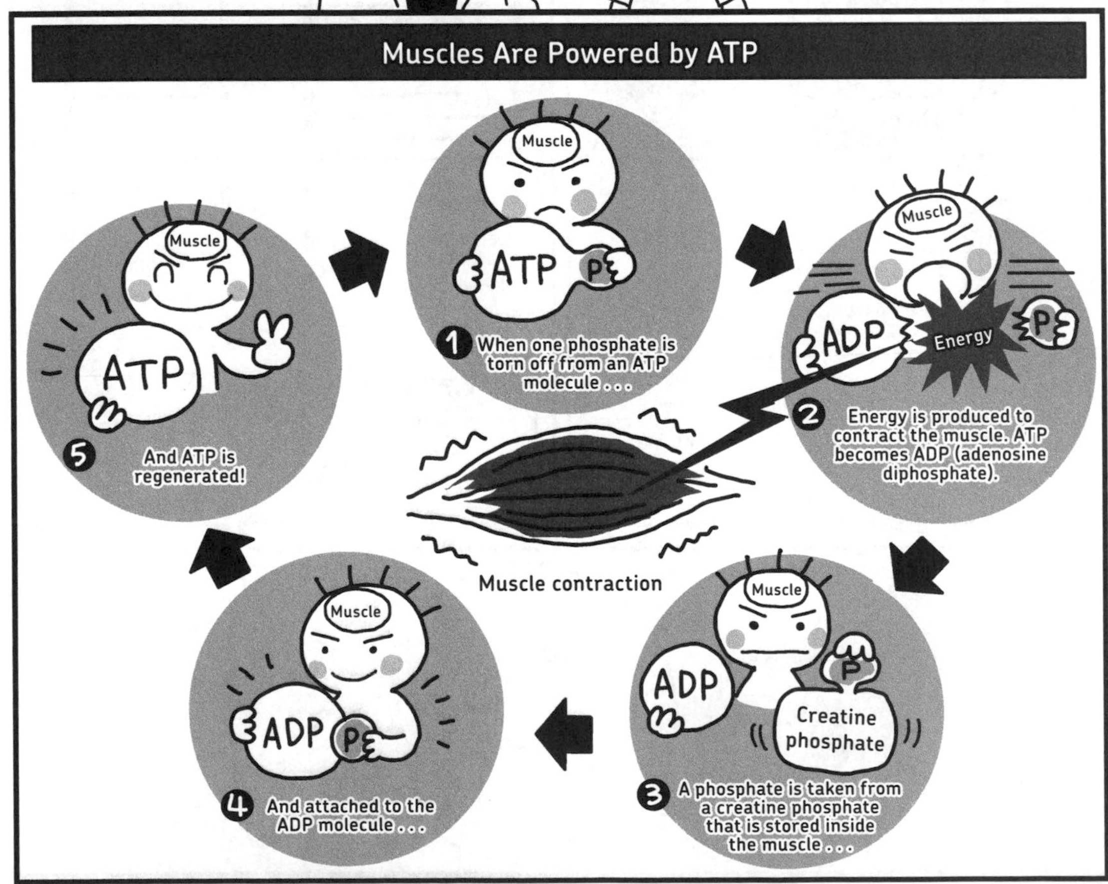

MY MUSCLES FEEL PERFECTLY STRETCHED AND RELAXED. I'M READY FOR A RUN!
GO FOR IT! WE CAN CONTINUE OUR CONVERSATION ON THE MOVE.
THESE GUYS COULD USE A WORKOUT ANYWAY.
HUFF
PUFF
SURE!
READY WHEN YOU ARE!
ALL RIGHT, SO DO YOU KNOW ABOUT RED AND WHITE FIBERS IN SKELETAL MUSCLES?
AH, I KNOW...
THE ONES WITH STAMINA ARE RED, AND THE ONES WITH EXPLOSIVE POWER ARE WHITE.

RED MUSCLE FIBERS (ALSO CALLED *SLOW-TWITCH FIBERS*) CONTAIN LOTS OF MYOGLOBIN,* WHICH RECEIVES AND STORES OXYGEN. RED MUSCLE CAN USE THAT OXYGEN TO GENERATE SUSTAINED ENERGY.

WHITE MUSCLE FIBERS (CALLED *FAST-TWITCH FIBERS*) HAVE MUCH LESS MYOGLOBIN, BUT THEY CAN CONTRACT IN SHORT, POWERFUL BURSTS USING ANAEROBIC ENERGY (ENERGY PRODUCED WITHOUT OXYGEN).

Red Muscle

- Migratory fish such as tuna have more red muscle for sustained swimming.
- Red muscle has lots of myoglobin.

White Muscle

- Fish with white muscle, such as sea bream, mostly swim in quick bursts.
- White muscle has little myoglobin.

* *MYOGLOBIN* IS A CHROMOPROTEIN THAT BINDS TO OXYGEN AND IRON. THESE GIVE IT A RED PIGMENT, MUCH LIKE HEMOGLOBIN IN THE BLOOD.

YOUR FRIENDS BACK THERE DON'T HAVE MUCH STAMINA.

THEIR MUSCLES MUST BE MOSTLY WHITE, DON'T YOU THINK?

HA! IN HUMANS, SKELETAL MUSCLES HAVE A MIX OF RED MUSCLE FIBERS WITH LOTS OF MYOGLOBIN AND WHITE MUSCLE FIBERS WITH LITTLE MYOGLOBIN.

THE PROPORTION DIFFERS FROM PERSON TO PERSON.

HEY, SLOW DOWN!

MS. KARADA, DON'T YOU THINK YOU'RE OVERDOING IT A LITTLE?
I NEED TO GO A BIT FARTHER...
TO FINISH MY INTERVAL.
WOBBLE...
HER FORM IS STARTING TO BREAK DOWN...
WHOA!
STAGGER
よろっ
TWEAK!
HEY, THAT LOOKED LIKE A BAD STEP—LET'S TAKE A BREAK.
OUCH OUCH!
I PUSHED TOO HARD.
I MUST HAVE GOTTEN CARRIED AWAY.

IS IT YOUR RIGHT KNEE THAT HURTS?
IT LOOKS LIKE YOU SPRAINED IT.
YOU SHOULD ICE YOUR KNEE AND REST.
HE'S SO CARING...

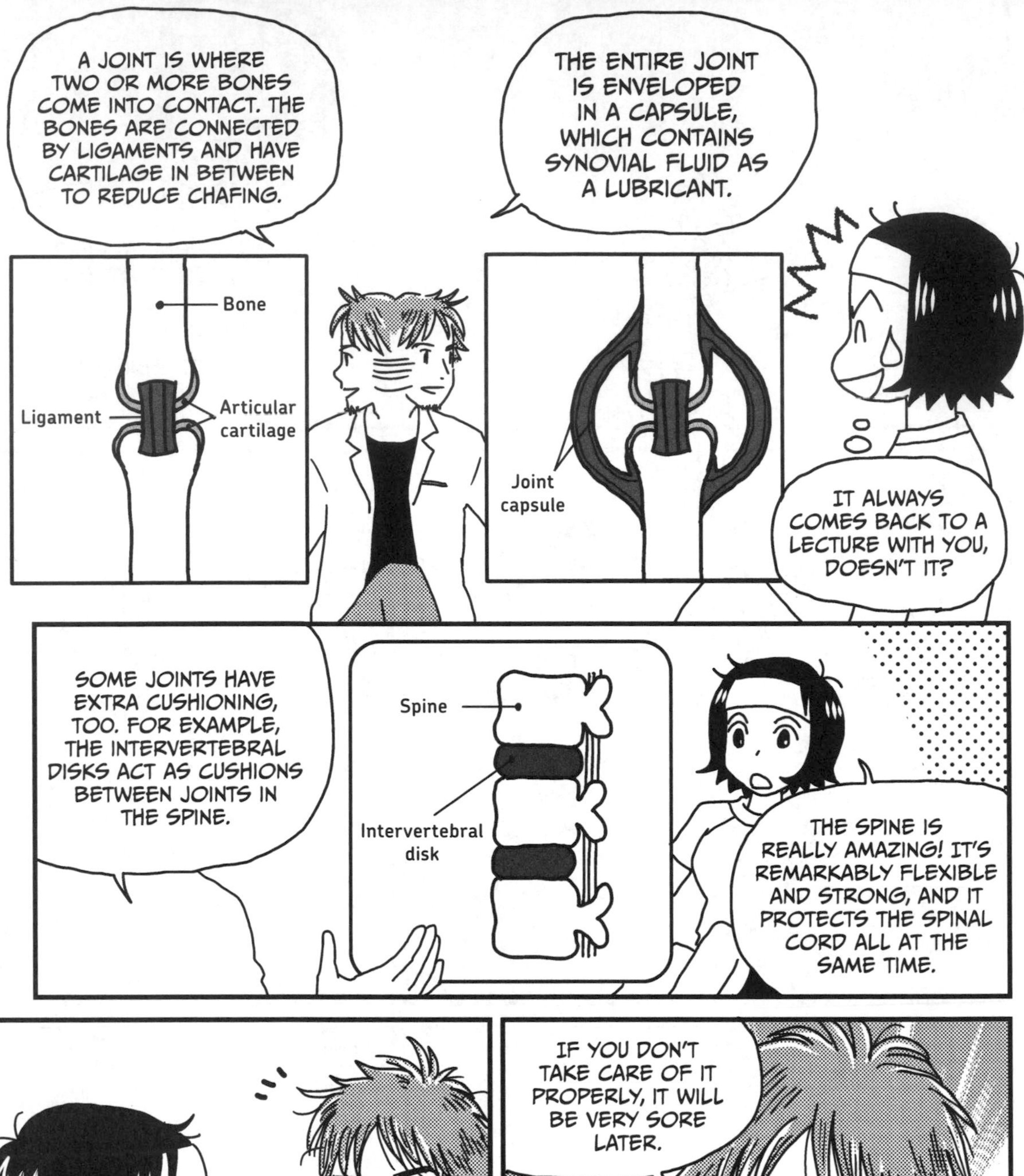
A JOINT IS WHERE TWO OR MORE BONES COME INTO CONTACT. THE BONES ARE CONNECTED BY LIGAMENTS AND HAVE CARTILAGE IN BETWEEN TO REDUCE CHAFING.
Bone
Ligament
Articular cartilage
THE ENTIRE JOINT IS ENVELOPED IN A CAPSULE, WHICH CONTAINS SYNOVIAL FLUID AS A LUBRICANT.
Joint capsule
IT ALWAYS COMES BACK TO A LECTURE WITH YOU, DOESN'T IT?
SOME JOINTS HAVE EXTRA CUSHIONING, TOO. FOR EXAMPLE, THE INTERVERTEBRAL DISKS ACT AS CUSHIONS BETWEEN JOINTS IN THE SPINE.
Spine
Intervertebral disk
THE SPINE IS REALLY AMAZING! IT'S REMARKABLY FLEXIBLE AND STRONG, AND IT PROTECTS THE SPINAL CORD ALL AT THE SAME TIME.

I THINK MY KNEE'S FEELING BETTER NOW.

IF YOU DON'T TAKE CARE OF IT PROPERLY, IT WILL BE VERY SORE LATER.

OKAY, I'LL REST IT. DIFFERENT KINDS OF JOINTS MOVE IN DIFFERENT WAYS, RIGHT?
MY SHOULDER CAN ROTATE ROUND AND ROUND, BUT MY KNEE CAN ONLY BEND AND STRETCH.
ROUND AND ROUND
BEND AND STRETCH
SO MUCH FOR TAKING A REST...
THAT'S RIGHT. A JOINT'S RANGE OF MOTION DEPENDS ON THE SHAPE OF THE BONES.
YOUR SHOULDER HAS A ROUND BALL JOINED TO A ROUND SOCKET, SO IT CAN ROTATE ROUND AND ROUND.
THIS TYPE OF JOINT IS CALLED A *BALL-AND-SOCKET JOINT.*
Ball-and-Socket Joint
THE KNEE IS MORE LIKE A DOOR HINGE: IT CAN FLEX AND EXTEND BUT NOT ROTATE.
SAME WITH OUR FINGER JOINTS.
THIS KIND OF JOINT IS CALLED A *HINGE JOINT.*
Hinge Joint
THESE ARE TWO SIMPLE KINDS OF JOINTS, BUT THERE ARE OTHER KINDS, TOO.
FINALLY, SOME TERMS THAT ARE NICE AND INTUITIVE.
THAT'S EASY ENOUGH.

THE MUSCLES THAT FLEX A HINGE JOINT ARE ALWAYS FOUND ON THE OPPOSITE SIDE OF THE MUSCLES THAT EXTEND THE JOINT.
OH MY!
Joint
FOR EXAMPLE, THE MUSCLE THAT BENDS THE ELBOW IS ATTACHED TO THE FRONT OF THE ARM, AND THE ONE THAT STRAIGHTENS THE ELBOW IS ATTACHED TO THE BACK.
THESE ARE CALLED THE *FLEXOR* AND *EXTENSOR* MUSCLES, RESPECTIVELY.
Antagonistic Muscles
WHOA...
Flexor muscle
Extensor muscle
MUSCLES THAT WORK IN OPPOSITE DIRECTIONS LIKE THIS ARE CALLED *MUTUALLY ANTAGONISTIC MUSCLES*. YOUR KNEE HAS THEM, TOO.

FLEXOR MUSCLES BEND MY RIGHT LEG IN...
Extensor muscle
BEND
Flexor muscle
AND EXTENSOR MUSCLES STRETCH IT OUT...
STRAIGHTEN
I FEEL LIKE I'M DOING THE HOKEY POKEY!

HEY, YOU REALLY OUGHT TO ICE THAT.
OUCH!
AND YOU SHOULD TAKE A BREAK FROM TRAINING TOMORROW!

YEAH, YEAH, I KNOW.

THAT REMINDS ME... TOMORROW IS YOUR FIRST ACTUAL LECTURE, ISN'T IT!
PRESSURE
I'M REALLY LOOKING FORWARD TO IT!

HMM. I WONDER IF HE'S OKAY.
GOTTA GO! BYE...
OW
NNNN
I HOPE I CAN STILL RUN THAT MARATHON...

EVEN MORE ABOUT MUSCLES AND BONES!

Let's talk a bit about how the body generates heat. Your body is constantly extracting energy from oxygen and food and burning that energy, releasing heat. Your body produces and radiates heat even when you're sleeping or sitting at your desk studying for a test. The more energy you use, the more heat is generated, so your body gets much warmer when you start exercising, as shown in Figure 8-1.

Figure 8-1: Your body generates heat even when you're sleeping, and the more active you are, the more heat it generates.

REGULATING BODY TEMPERATURE

You've probably noticed that when you exercise, your body heats up. That's because heat is produced when skeletal muscles contract. When it's cold, your muscles contract rapidly and you shiver (see Figure 8-2). This produces heat that the blood carries throughout the body to maintain the body's temperature.

Skeletal muscles are the main producers of heat in the body, but they aren't the only ones. The digestive system, heart, brain, and liver all produce heat as well—in fact, any active cell produces heat to some extent. The body gets warmer after a meal, not only if the food is warm, but also because of the increased activity of the digestive system.

Humans are *homeotherms*, which means our body temperature has to stay within a certain range. If our body temperature is too high or too low, our bodies can't function properly.

DID YOU KNOW?

A special kind of fat tissue called *brown fat* consists of cells with extra mitochondria, plus a particular type of protein that affects the production of ATP. The result is that as brown fat consumes calories, it produces heat instead of ATP. This type of tissue is particularly common in newborns and hibernating animals.

Figure 8-2: We shiver when we're cold. These rapid muscle contractions produce heat!

Our bodies have to keep from overheating too, right? Isn't that what perspiration is for, to help disperse heat?

That's right. Your body temperature must be maintained at approximately 96.8–100.4 degrees Fahrenheit (36–38 degrees Celsius). The body cools as heat is dispersed. Heat can leave the body by escaping through the skin, by respiration, and by perspiration (see Figure 8-3). When you exercise vigorously and start heating up, your body starts perspiring more to cool down.

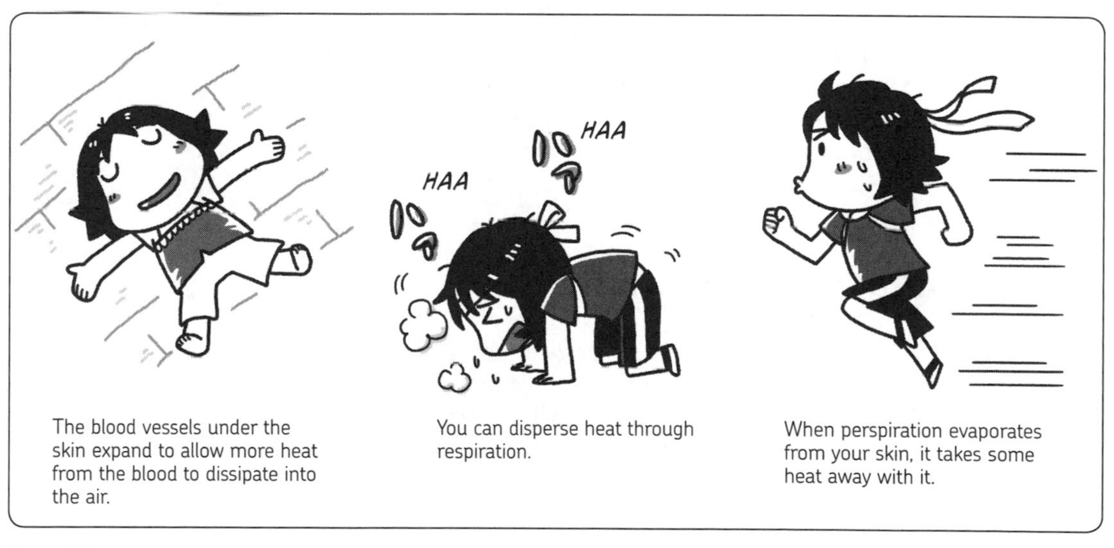

Figure 8-3: Methods of dissipating body heat

When it's hot outside, your skin flushes because the blood vessels under your skin expand. This allows more blood to flow out to the surface of the skin so that the heat in your blood can radiate from your skin to the air around you. Perspiration also cools your body through a process called *evaporative cooling*. In fact, when the air temperature is higher than your body temperature, sweating is the only way to eliminate that body heat. Also, sweat glands, like muscles, can get larger and more productive if you use them more often, so exercising regularly will help your body adapt better to the summer heat! It's important to stay hydrated when it's hot out, because your body loses a lot of liquid as it sweats.

Do you know why the body can't function at a temperature higher than 107.6 degrees Fahrenheit (42 degrees Celsius)?

Hmm . . . why is that?

If the body temperature exceeds 107.6 degrees Fahrenheit, the proteins in the body begin to degenerate, and the body can no longer function. (In fact, old mercury thermometers often only go up to this temperature!)

The body's temperature is regulated by the thermoregulatory center of the hypothalamus. The hypothalamus uses several mechanisms to maintain body temperature. When you're too cold, the posterior portion of the hypothalamus coordinates responses such as shivering, controlling blood flow to the skin, and secreting hormones like norepinephrine and epinephrine. When you're too hot, the anterior hypothalamus coordinates the opposite responses.

DID YOU KNOW?

The carotid artery, axillary artery, and femoral artery are three thick arteries that run near the surface of the body in the neck, armpit, and groin, respectively. Cooling or warming these areas can have a strong impact on body temperature by changing the temperature of the blood as it flows through the arteries.

BONES AND BONE METABOLISM

The main role of the approximately 206 bones in the human body is structural support. If there were no bones, the body would collapse, and it wouldn't be able to move. But this is not the only role that bones play. They also store calcium, and they contain bone marrow, which creates new blood cells.

Bone Marrow

Bone marrow is found inside our bones. If you gathered all the bone marrow in a single human body together, it would be about the same size as the liver. Marrow is full of *progenitor cells*, which eventually specialize into specific types of blood cells, like red blood cells, white blood cells, and platelets. This ability of progenitor cells to morph into whatever specific cells the body needs is an active area of interest to stem cell researchers who want to find a way to treat disease with unspecialized, multipurpose cells that can regenerate damaged organs.

Blood cells are constantly being created in the marrow of flat bones like the pelvis and sternum, but long and narrow bones, like those in the legs and arms, don't produce many blood cells. In fact, long bones only produce blood until around 20 years of age. After that time, the marrow becomes inactive and loses its red coloration, and the red marrow becomes yellow marrow.

When a fetus is developing in the womb, its blood cells are created in the liver and spleen. These organs are able to resume the production of blood cells later in life if bone marrow cannot produce enough blood.

Bones are designed to be both strong and lightweight. Many bones use an internal meshwork structure called *trabecular bone*, which looks like a sponge. Trabecular bone (also called *cancellous bone*) is often found at the ends of long bones and in bones with complicated shapes like the shoulder blade (scapula), vertebrae, and pelvis. The strong exterior of bones is made of *cortical bone*, which is much more compact. In the case of a long bone like the femur, the shaft of the bone is cortical bone with a hollow center (called a *medullary cavity*). Cortical and trabecular bone work together to give bones their incredible strength. The types of bone are illustrated in Figure 8-4.

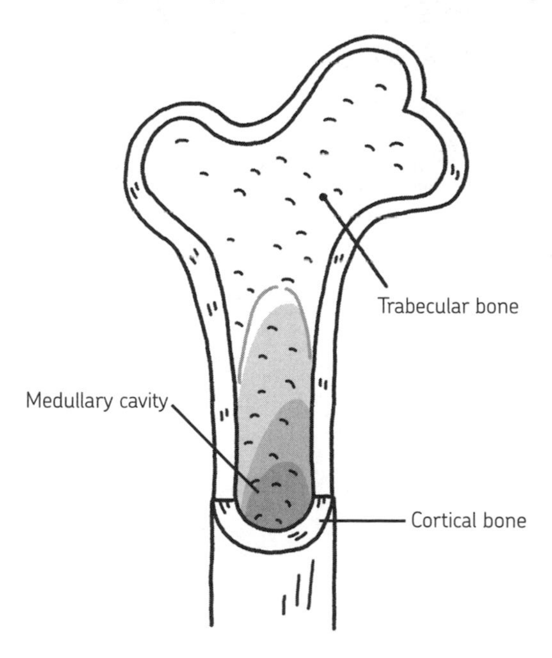

Figure 8-4: Basic structure of bone

At the microscopic level, bone is made up of a framework of a protein called *collagen* (pound for pound, it's as strong as steel). Collagen becomes calcified in such a way that it has flexibility instead of being totally rigid. This ability to bend without snapping helps your bones withstand extremely powerful external forces.

Who knew bones were so complex! Bone is always replacing itself too, isn't it?

That's right. Even after you stop growing, your bones are always dissolving a little at a time (*resorption*), and new bone is being reformed there (*ossification*). It's said that by the time you're about two years old, all the original bones in your body have been replaced.

Bones are dissolved and regenerated by cells called osteoclasts and osteoblasts. *Osteoclasts* dissolve and break down the bone structure little by little. *Osteoblasts* regenerate the bone by attaching calcium. This whole process is called *bone metabolism*, which both enables and prunes bone growth, always keeping the microscopic structure in order (see Figure 8-5).

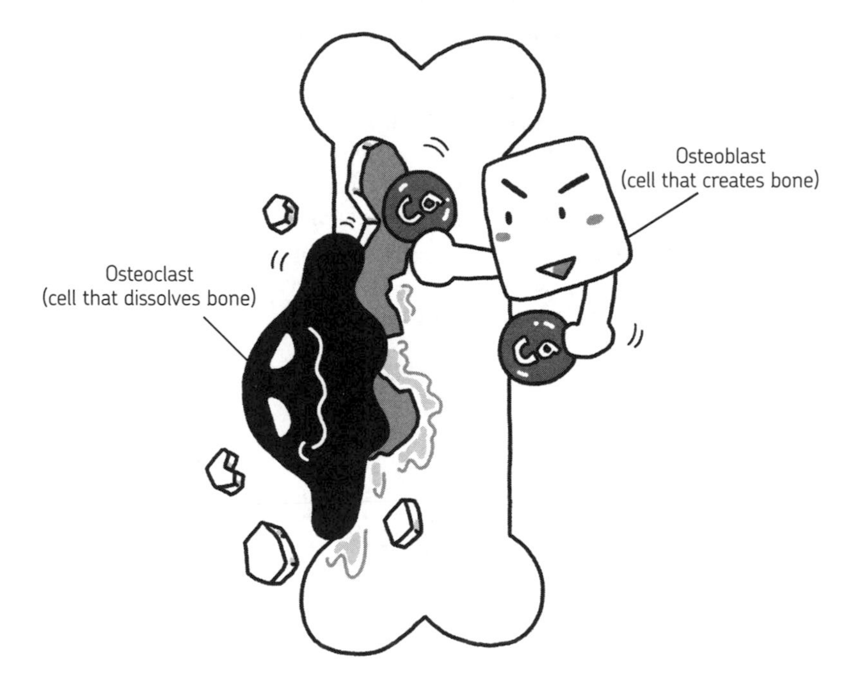

Figure 8-5: Osteoclasts and osteoblasts at work

DID YOU KNOW?

Bone contains 99 percent of the calcium in the human body. Bone metabolism has a close relationship with estrogen, which slows the resorption of bone. As a result, after menopause—when women produce significantly less estrogen—women are vulnerable to dangerous bone loss called *osteoporosis*, which can leave bones brittle.

Of course, when osteoclasts break down bone, they aren't just being a nuisance. They remove old bone that has developed microscopic cracks over time. They also extract calcium from bones in order to maintain the correct concentration of calcium in your blood.

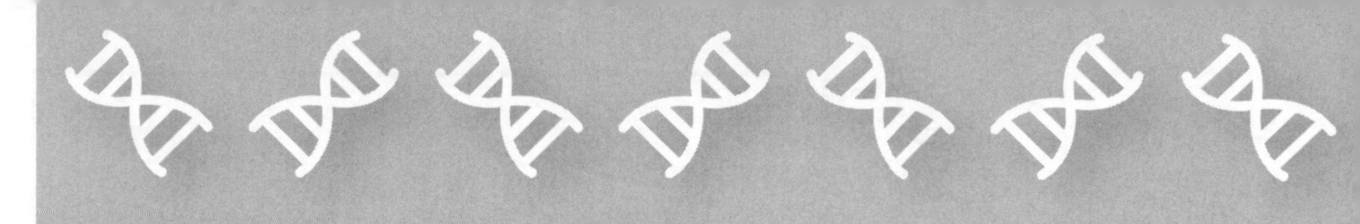

9 CELLS, GENES, AND REPRODUCTION

STORING AND REPLICATING BIOLOGICAL BLUEPRINTS

GREAT TURNOUT, HUH?
Y-YES...
WHOA, SHE SPOKE TO ME. THAT'S A FIRST.
THEY ALL CAME OUT FOR JUNIOR'S FIRST LECTURE.
HUH? WHAT'S UP WITH THIS "JUNIOR" STUFF AGAIN? THE DORM MOTHER WAS ALSO CALLING HIM "JUNIOR."
KOUJO MEDICAL SCHOOL
New Department of Sports and Health Science
Open Campus August 3
Featuring a Q&A with star athletes!
ぴっ
STAR ATHLETES??
THAT GUY OVER THERE IS SATO, THE WRESTLER!
AND HE'S MIZUSHIMA, THE SWIMMER!!
QUIET!
SHHHH

CELLS ARE THE SMALLEST UNITS OF LIFE AND ARE OFTEN CALLED THE BUILDING BLOCKS OF THE HUMAN BODY.
OUR BODIES CONTAIN TRILLIONS OF CELLS.
DIFFERENT KINDS OF CELLS HAVE DIFFERENT SPECIALIZED FUNCTIONS, BUT BY AND LARGE THEY ALL SHARE A BASIC STRUCTURE AS WELL AS BASIC TASKS LIKE CONVERTING NUTRIENTS INTO ENERGY.
Cell
GREAT, HE'S COVERING CELLS!
BASIC STRUCTURE OF THE CELL
THE CELL MEMBRANE IS A SEMIPERMEABLE WALL THAT SURROUNDS THE CELL. THE LIQUID THAT FILLS THE CELL IS CALLED *CYTOPLASM*.
Ribosome
Golgi apparatus
Mitochondria
Nucleus
Cell membrane
Cytoplasm
THE CELL ALSO CONTAINS ORGANELLES, WHICH EACH HAVE A UNIQUE ROLE.
MITOCHONDRIA PRODUCE ATP, RIBOSOMES HELP ASSEMBLE PROTEINS, AND THE GOLGI APPARATUS STORES AND SECRETES PROTEINS.
THE NUCLEUS STORES GENETIC INFORMATION.
AM I GOING INTO TOO MUCH MICROSCOPIC DETAIL FOR YOU?

WHEN IT'S TIME TO BUILD A PARTICULAR CAR, THE BLUEPRINT IS COPIED FROM THE FILES AND USED TO ASSEMBLE THAT CAR.

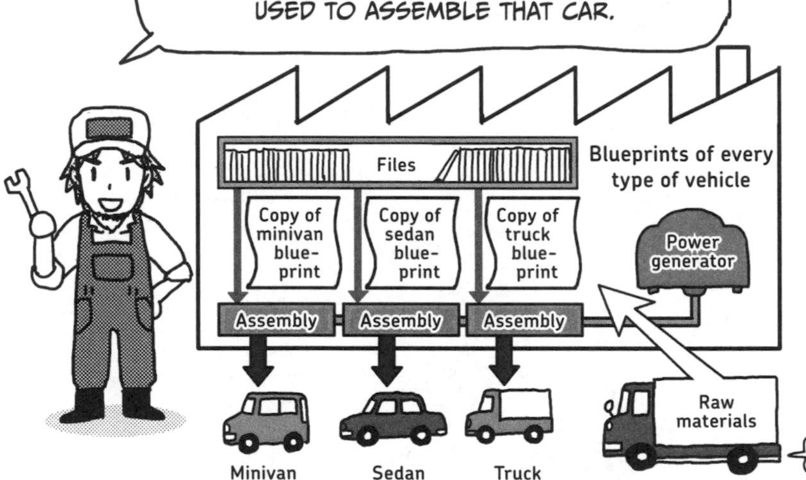

IN A CELL, A SECTION OF DNA IS COPIED TO A STRAND OF RNA, AND RIBOSOMES SYNTHESIZE PROTEINS ACCORDING TO THAT RNA COPY.

Mitochondria supply ATP as energy ("ATP and the Citric Acid Cycle" on page 74).

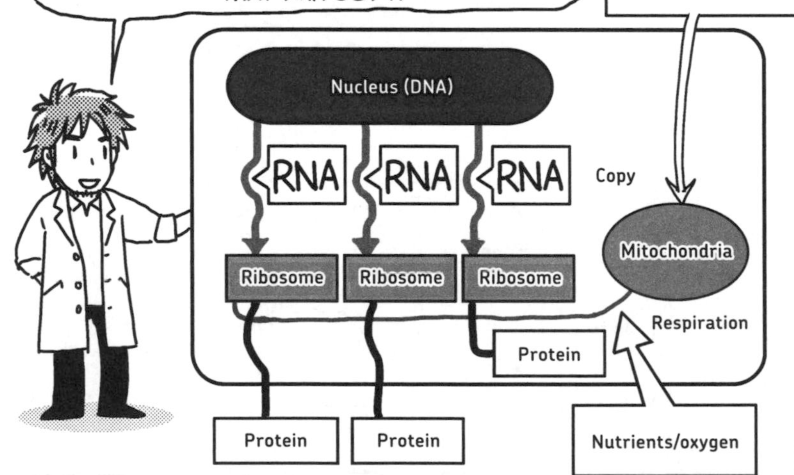

HA HA HA
THERE'S JUST ONE DIFFERENCE—EVEN MY MUSCLE CELLS HAVE MUSCLE CELLS.
SERIOUSLY, THOUGH, LISTEN TO JUNIOR AND YOU'LL LEARN A THING OR TWO.
UH, THANKS!
EVEN THOUGH CELLS SHARE A BASIC STRUCTURE, THEY CAN STILL BE PRETTY DIFFERENT. DID YOU KNOW THAT NEURONS HARDLY DUPLICATE OR DIVIDE AT ALL AFTER YOU ARE BORN?
BUT OTHER CELLS, SUCH AS SKIN CELLS, ARE CONSTANTLY REPLENISHED—OLDER CELLS DIE AND NEW CELLS ARE CREATED TO REPLACE THEM.
HEY...
HOW'D HE GET ALL THOSE ALL-STAR ATHLETES TO SHOW UP?
WHISPER
THOSE GUYS ARE FORMER STUDENTS OF THE PROFESSOR'S FATHER.
ボソボソ
THE PROFESSOR HAS BEEN PALLING AROUND WITH THEM SINCE HE WAS A KID.
OH, HE'S FROM AN ACADEMIC FAMILY?

EPITHELIAL TISSUE LINES THE SURFACES OF YOUR BODY. IT FORMS THE OUTER LAYER OF SKIN AND THE LINING OF YOUR ORGANS. THESE CELLS FORM PROTECTIVE SHEETS THAT CAN ABSORB AND SECRETE FLUIDS.

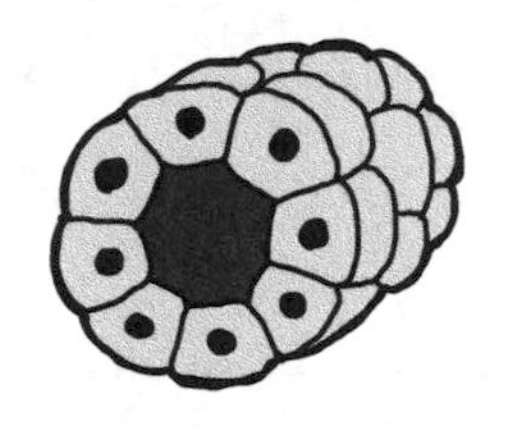

CONNECTIVE TISSUE PROVIDES STRUCTURE THROUGHOUT THE BODY. IT'S MOSTLY FOUND IN CARTILAGE, BONE, AND FAT.

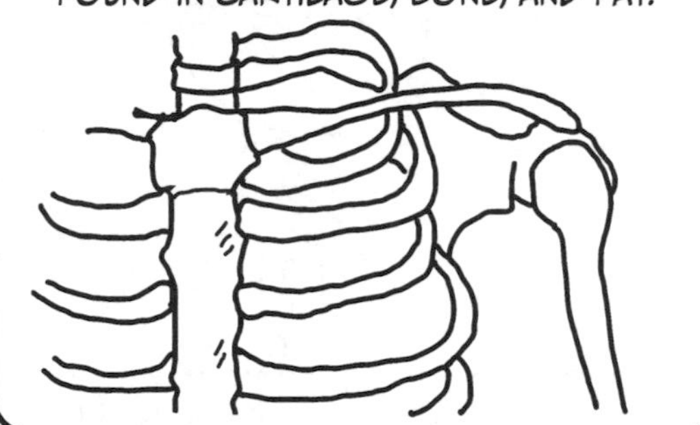

MUSCLE TISSUE CELLS ARE DESIGNED TO CONTRACT. THERE ARE THREE TYPES, SHOWN BELOW:

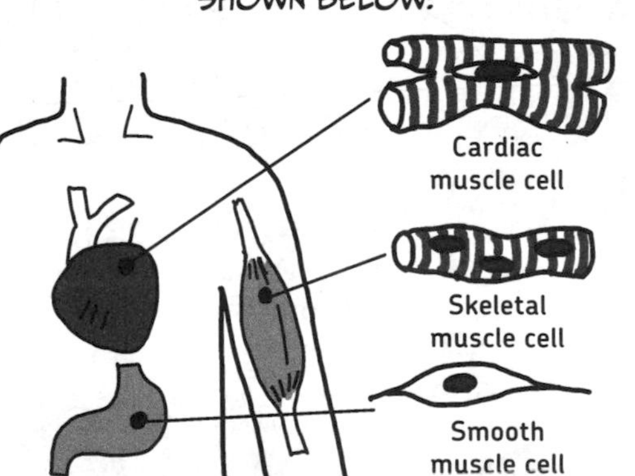

NERVOUS TISSUE CONSISTS OF NEURONS AND GLIAL CELLS, WHICH SUPPORT NEURONS AND PROVIDE THEM WITH NUTRIENTS.

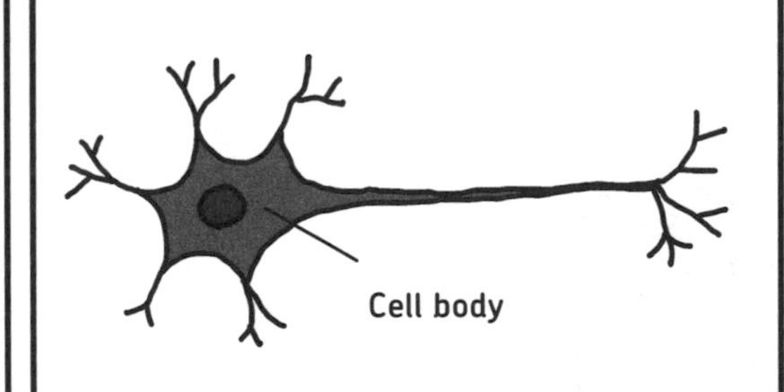

SO CELLS COLLECTED TOGETHER BECOME TISSUE, AND VARIOUS KINDS OF TISSUE ARE COMBINED TO FORM ORGANS...

AND ALL THOSE ORGANS ADD UP TO MAKE A COMPLETE HUMAN BODY.

GENES AND DNA

WHAT'S THE DIFFERENCE BETWEEN DNA AND GENES?

OOH, I WANTED TO ASK THAT TOO.

OUR GUEST HAS BECOME A STUDENT...
ACTUALLY, THAT'S A COMMON QUESTION.

DNA REFERS TO THE COMPLETE, DOUBLE-HELIX STRUCTURE...
THAT CONTAINS OUR ENTIRE BIOLOGICAL BLUEPRINT.

GENES REFER TO A PARTICULAR STRIP OF INSTRUCTIONS THAT CONTROL A PARTICULAR TRAIT. IT'S SORT OF LIKE DNA IS A LONG STRIP OF PAPER AND GENES ARE PARTICULAR SETS OF INSTRUCTIONS ON THAT STRIP.
The long connected strip of paper is the DNA.
The individual instructions are genes.
IN FACT, ONLY A TINY PERCENTAGE OF THE ENTIRE HUMAN GENOME CONSISTS OF GENES—THE VAST MAJORITY OF DNA DOES NOT DIRECTLY CONTROL ANY TRAITS OR PROTEINS.

* A JAPANESE PROVERB SIMILAR TO "LIKE FATHER, LIKE SON."

AHEM, SO I WAS SAYING...
COUGH
EVEN IN THE SPORTS AND HEALTH SCIENCE DEPARTMENT, WE PURSUE GENETIC RESEARCH.
A
4
C
9
10
D
13
14
15
E
16
17
18
G
21
22
X
Y
FOR EXAMPLE, WE INVESTIGATE THE DNA OF FORMER OLYMPIC ATHLETES...
TO LOOK FOR A GENETIC BASIS FOR CHARACTERISTICS SUCH AS "EXPLOSIVENESS" OR "ENDURANCE."
IT WOULD BE WONDERFUL IF WE COULD ESTABLISH BETTER TRAINING METHODS OR NUTRITIONAL SUPPLEMENTS BASED ON RESULTS FROM GENETIC RESEARCH.
I'D PARTICIPATE IN ANY RESEARCH STUDY FOR YOU, OSAMU!
FLEX
BUT IT'S NOT ALL GENETICS. I WORKED LONG AND HARD TO BE A TOP WRESTLER!
GET OUT OF HERE.
I'M SORRY.

ISN'T THERE SOME TRUTH TO THAT?
GENES CONTROL CERTAIN TRAITS, BUT AREN'T THERE MANY OTHER FACTORS THAT DETERMINE YOUR ABILITY AS AN ATHLETE, LIKE TRAINING TECHNIQUES AND HARD WORK?

OF COURSE.
IN FACT, I'D SAY THE SAME APPLIES TO ACADEMIC APTITUDE.

HUH?

I ALWAYS WONDERED ABOUT THAT.
MY PARENTS WERE BOTH GOOD STUDENTS. BUT I THINK THERE'S MORE TO IT.
THEY ENCOURAGED ME TO STUDY HARD.

WELL, THIS CONCLUDES THE DEPARTMENT OF SPORTS AND HEALTH SCIENCE OPEN CAMPUS LECTURE FOR THE COURSE THAT WILL BE OFFERED NEXT TERM.
パチパチ
パチパチ

I CAN'T BELIEVE I DIDN'T KNOW THE PROFESSOR IS THE SON OF THE UNIVERSITY PRESIDENT.
I WONDER IF HIS TEACHING SKILLS ARE GENETIC...
CLAP CLAP
CLAP CLAP

EVEN MORE ABOUT CELLS, GENES, AND REPRODUCTION!

Heredity is a phenomenon in which characteristics and traits are passed on from cell to cell or from parent to child. In this section, we'll talk about the two mechanisms that allow genes to be passed on to new cells or offspring: cell division and reproduction.

CELL DIVISION

As cells in the skin, bone, and certain other parts of the body degenerate, new replacement cells are created by *cell division*. Cell division is also the process that allows a single, fertilized egg cell to grow into a baby during pregnancy (which we'll come back to a bit later).

CHROMOSOMES

Before we talk about cell division, though, we have to learn a bit about chromosomes. *Chromosomes* are very long strands of DNA wrapped around proteins. You can think of them as bundles of DNA packaged neatly to make them more manageable in the cell. Humans have 46 chromosomes in 23 pairs. Having pairs ensures that you have two copies of all the essential DNA your body needs. Cells that have two copies of each chromosome are called *diploid*.

Of the 23 pairs, one pair (two chromosomes) are special *sex chromosomes* (also called *allosomes*), while the other 22 pairs (44 chromosomes) are called *autosomes*. In females, both sex chromosomes are X chromosomes, while males have one X chromosome and one Y chromosome.

DID YOU KNOW?

The number of chromosomes differs according to the type of organism. Humans have 46 chromosomes, but dogs have 78 and fruit flies have only 8.

Usually DNA is mostly unwound and scattered all over the nucleus, but when a cell is preparing to divide it packages up each strand of DNA neatly into chromosomes. At that point, we can use a microscope to look at the shape of the 46 chromosomes and tell whether the cell has two Xs or an X and a Y.

Congratulations, It's a . . .

The sex chromosome of a sperm can be either X or Y, but the sex chromosome of an egg is always X. That means that when an egg is fertilized, the sex chromosome of the sperm determines the sex of the baby—if the sperm has an X chromosome, the child will be female, and if it has a Y chromosome, the child will be male.

MITOSIS

Hmm, so we can only see the shape of the chromosomes during cell division.

That's right. Normally DNA is arranged in long, thin coils so that the cell can read and use the DNA code, but these coils are bundled up into chromosomes during a type of cell division called *mitosis*. Let's walk through that process.

First, before mitosis begins, DNA is replicated in the nucleus. Then the nuclear membrane disintegrates, and DNA bundles up to form chromosomes shaped like the letter *X*. Each *X* contains two strands of identical code, the original and a duplicate, which are side by side and held together at the middle. Those duplicate sets of chromosomes are then lined up in the center of the cell, and each set of duplicate chromosomes is pulled apart by threadlike structures called *microtubules* so that each cell will get only one copy. Finally, the center of the cell squeezes tight until the cell splits. The end result is two identical cells, as shown in Figure 9-1.

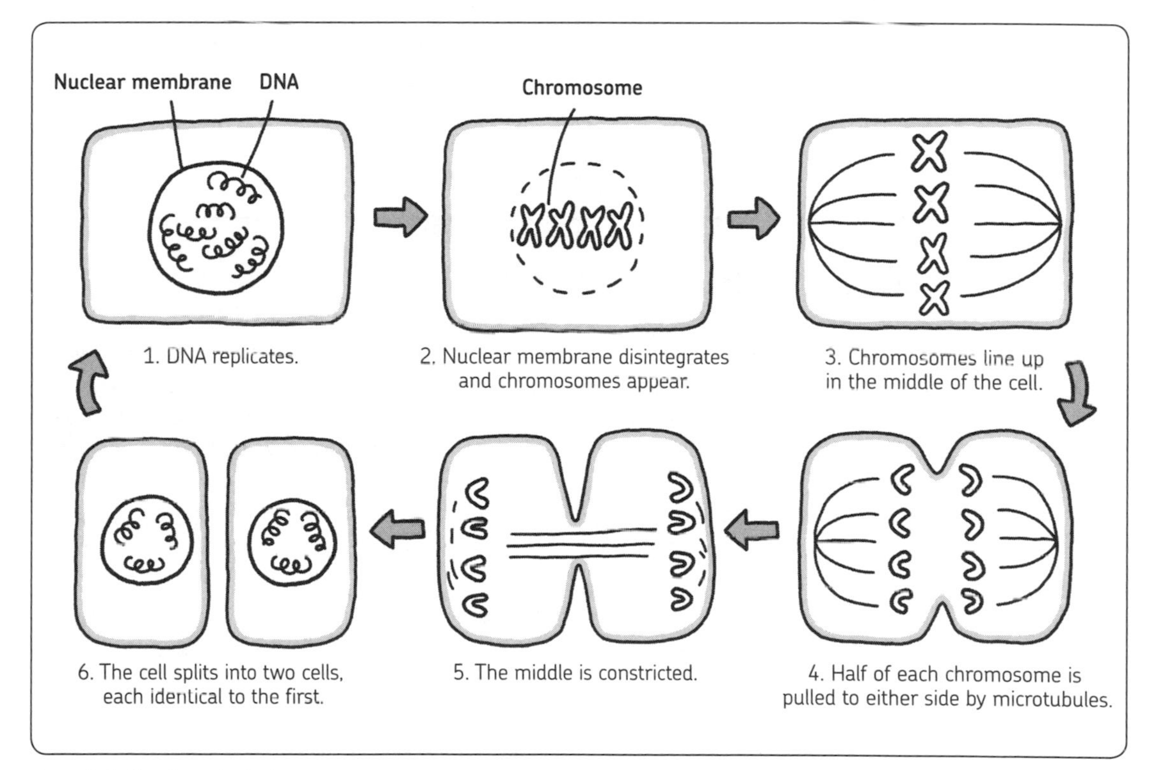

Figure 9-1: Flowchart of mitosis

MEIOSIS

You know that reproductive cells (also called *germ cells* or *gametes*) undergo special cell division, right?

Right, because egg and sperm cells only need half the chromosomes of a regular body cell. Normal cell division produces cells with the same number of chromosomes as the original cell. Germ cells, however, have half as many chromosomes as a normal cell and are produced through a special kind of cell division called *meiosis*.

Meiosis begins with a cell that has duplicated all its DNA, just as in mitosis. But then, instead of dividing just once, the cell divides twice. The result is four germ cells, each with 23 chromosomes. These germ cells are *haploid* cells, as opposed to diploid cells, which have 23 *pairs* of chromosomes.

> **Interphase**
>
> So far we've been talking about how cells divide, called the *division phase*, but most of the time cells are in a stage of *interphase*. In interphase, cells use the information in DNA to grow, collect nutrients, make proteins, and perform their specialized functions in the body. Whenever necessary, a cell in interphase can initiate another division phase to create two cells, which then start in interphase. Together, interphase and the division phase make up the complete cell cycle.

SEXUAL REPRODUCTION

In order to reproduce, single-celled organisms like bacteria or algae simply divide in two. With this type of reproduction, called *asexual* reproduction, the new, independent organism is exactly the same as the parent. By contrast, almost all multicellular organisms (including humans) rely on sexual reproduction, which produces offspring different from either of its parents.

Sexual reproduction depends on gametes: the sperm and egg cells. Sperm cells are produced in the testicles of a male, while egg cells are produced and matured in the ovaries of a female.

The testicles produce new sperm all the time, but the number of eggs in the ovaries is finite, isn't it?

Females are born with one to two million *ovarian follicles*. Each ovarian follicle is a group of cells that surround a single, immature egg cell (also called an *ovum*). After the onset of puberty, each month a single ovarian follicle matures and becomes ready for fertilization, while several thousand other ovarian follicles are steadily reabsorbed by the body. Eventually the supply of ovarian follicles runs out.

When an ovarian follicle matures, the egg cell is released from the follicle and pushed out of the ovary. This process is called *ovulation* (see Figure 9-2). The egg cell is then guided by the *fimbriae*, a kind of fringe of tissue, into the *fallopian tube* where it might encounter sperm. Meanwhile, the ovarian follicle that produced the egg changes shape and becomes a *corpus luteum*. If the egg cell encounters a sperm cell, the egg is fertilized and will then send a message to the corpus luteum, which will continue to play a crucial role by secreting important hormones to promote pregnancy (see "Sex Hormones" on page 219).

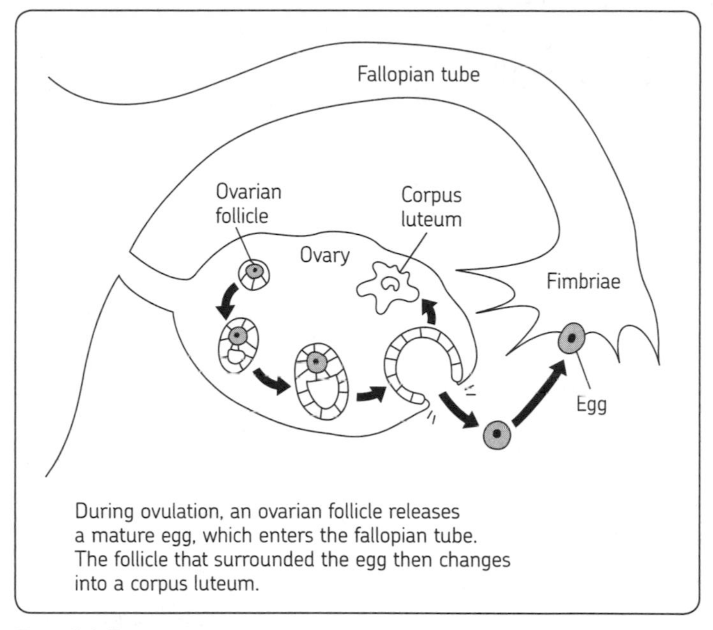

Figure 9-2: The ovulation process

Now let's turn back to the sperm. When sperm cells enter the vagina, a vigorous competition to reach the egg ensues (see Figure 9-3).

Only one sperm can fertilize the egg, right? How many sperm are usually competing?

The number of sperm ejaculated at one time ranges from tens of millions to a hundred million. During the competition, some sperm power ahead while others lose their way and drop out. The number of sperm that typically reach the fallopian tubes is estimated to be in the tens of thousands, and about 100 manage to arrive in the vicinity of the egg. Only one single sperm is capable of ultimately fertilizing the egg, but it's entirely possible that none will make it.

DID YOU KNOW?

The lifespan of a mature egg is between half a day and one day, and the life span of a sperm is approximately two to three days.

Between 10 and 100 million sperm start at the same time.

The sperm race to the womb. Some get lost or die along the way.

Sperm continue to push on after deciding whether to go left or right at the entrance to the fallopian tubes.

Relatively few sperm reach the vicinity of the egg. The very first sperm to reach the egg will fertilize it.

Figure 9-3: The fertilization race

Cilia along the inner surface of the fallopian tube carry the egg to the uterus, and if it's been fertilized, the egg begins many rounds of cell division and starts to grow. The fertilized egg then implants itself wherever it lands on the *endometrium* (the inner wall of the uterus), and that's where it will develop throughout pregnancy, as shown in Figure 9-4.

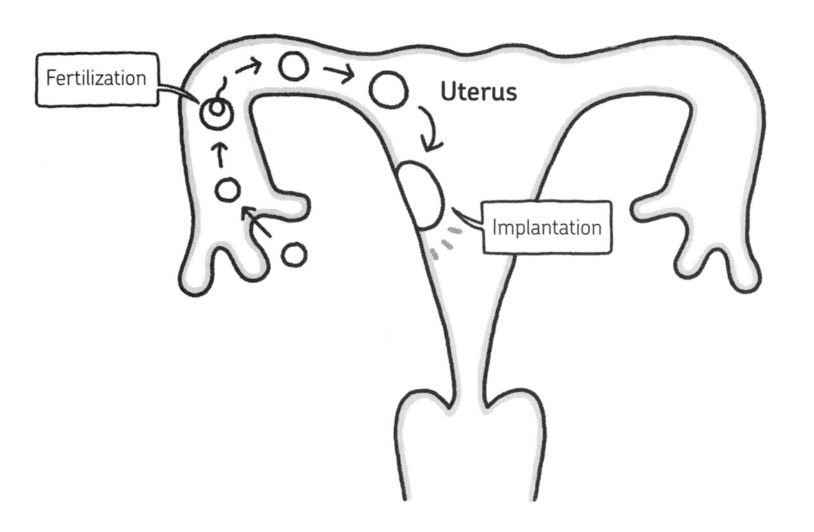

Figure 9-4: From fertilization to implantation

But cell division starts even before the fertilized egg reaches the uterus, right?

That's right. Cell division begins immediately after fertilization, and then the cells start to specialize around the time of implantation in the uterus.

Gestation Period

From the time of fertilization, it takes approximately 38 weeks for the fertilized egg to develop into a baby capable of surviving outside the womb. But the pregnancy period is often considered to begin on the first day of the last menstrual period, which is roughly two weeks before ovulation and fertilization occur. That makes for a total timeline of 40 weeks.

10 THE ENDOCRINE SYSTEM

SENDING SIGNALS THROUGH THE BLOODSTREAM

WHAT IS THE ENDOCRINE SYSTEM?

Main Human Endocrine Organs

Hypothalamus and Pituitary Gland

The hypothalamus links the nervous system and the endocrine system, and keeps the body's temperature and chemistry in balance. The pituitary gland is like a control center—it signals other glands to secrete their hormones.

Thyroid Gland

This butterfly-shaped gland in the neck secretes hormones that regulate metabolism. The parathyroid glands underneath it regulate calcium levels and bone structure.

Adrenal Glands

These triangular glands sit on top of the kidneys. They are divided into the adrenal medulla at the center and the adrenal cortex surrounding it.

Pancreas

Besides secreting digestive fluids (pancreatic juice), the pancreas also secretes hormones that regulate the uptake and distribution of nutrients throughout the body.

Ovaries (female)

The ovaries secrete estrogen and progesterone and produce ova (egg cells).

Testicles (male)

The testicles secrete androgens (male hormones) and produce sperm.

Information slowly filters out to various targets.

Information is sent quickly from point to point.

WAIT, BUT HOW DO HORMONES SEND SIGNALS TO SPECIFIC ORGANS...
IF THEY ARE JUST FLOATING IN THE BLOODSTREAM?
MAN, YOU REALLY DO HAVE SOME STUDYING LEFT TO DO. LET'S GO A LITTLE DEEPER.
A HORMONE AFFECTS ONLY CERTAIN TARGET INTERNAL ORGANS.
THAT'S BECAUSE ORGAN CELLS HAVE RECEPTORS THAT RESPOND ONLY TO CERTAIN HORMONES.
IT'S SORT OF LIKE A KEYHOLE. A HORMONE ACTS ON AN ORGAN ONLY IF ITS KEY FITS THAT ORGAN'S KEYHOLE.
Blood vessel
Target organ
THE KEY FITS!
Instruction
THE KEY FITS!
Instruction
Target organ
I SEE!
ALTHOUGH THE HORMONE PASSES BY LOTS OF CELLS, ONLY CERTAIN CELLS CAN RECEIVE THE HORMONE'S INSTRUCTIONS.
A SINGLE HORMONE CAN AFFECT MULTIPLE ORGANS, INCLUDING OTHER ENDOCRINE GLANDS. AND EVEN TRACE AMOUNTS OF A HORMONE CAN HAVE A SIGNIFICANT EFFECT.
WOW, IT'S A REALLY INTRICATE SYSTEM.
WHAT HAPPENS IF THERE'S A HORMONE IMBALANCE?
IT DEPENDS— IT COULD AFFECT ANYTHING FROM MOOD TO BONE GROWTH TO APPETITE.

I KNOW! I KNOW!

THAT'S HOW HORMONE LEVELS ARE ADJUSTED IN THE BLOOD. IF THE LEVELS GET TOO HIGH OR TOO LOW...

THE GLANDS ADJUST THEIR SECRETION LEVEL.

YES, THAT'S RIGHT. AS AN EXAMPLE...

LOOK AT THE THYROID GLAND HORMONES.

BALANCING HORMONE LEVELS

SAY THE HYPOTHALAMUS AND PITUITARY GLAND DETECT THAT THE CONCENTRATION OF THYROID HORMONES IS TOO LOW. IN RESPONSE, THEY SECRETE HORMONES THAT STIMULATE THYROID HORMONES.

INCREASE!

Upstream → Downstream

Hypothalamus hormones → Pituitary gland hormones → Thyroid gland hormones

As the upstream hormones increase, the downstream hormones increase in response.

LET'S CALL THE HYPOTHALAMUS AND PITUITARY HORMONES UPSTREAM SINCE THEY COME EARLIER IN THE CAUSAL CHAIN.

I SEE, THEY BOOST THYROID HORMONE SECRETION FURTHER DOWNSTREAM.

THAT'S CORRECT.

BUT WHEN THE THYROID HORMONE LEVEL GETS TOO HIGH, NEGATIVE FEEDBACK KICKS IN: THE HYPOTHALAMUS AND PITUITARY GLAND DETECT THE OVERABUNDANCE AND REDUCE THEIR SECRETION OF THYROID-STIMULATING HORMONES...

AND THE VOLUME OF THYROID GLAND HORMONES DECREASES IN RESPONSE.

From the pituitary gland

From the thyroid gland

Up-stream

Down-stream

THIS CONSTANT FEEDBACK KEEPS HORMONE LEVELS IN BALANCE. THE LEVEL OF HORMONES FROM UPSTREAM SOURCES LIKE THE HYPOTHALAMUS CONTROLS HORMONE SECRETION DOWNSTREAM IN PLACES LIKE THE THYROID.

BUT AT THE SAME TIME, THROUGH NEGATIVE FEEDBACK, THE HORMONES FROM DOWNSTREAM AFFECT THE HORMONES UPSTREAM.

DECREASE!

Upstream ← Downstream

Hypothalamus hormones ← Pituitary gland hormones ← Thyroid gland hormones

Negative feedback

IN THIS CASE, HIGH THYROID HORMONE LEVELS TRIGGER THE UPSTREAM GLANDS TO REDUCE SECRETION.

IT'S LIKE HORMONE PRODUCTION CONTINUALLY ACCELERATES AND BRAKES TO STAY NEAR A CONSTANT LEVEL.

IT'S ALSO TRUE THAT SEVERAL DIFFERENT HORMONES CAN HAVE THE SAME KIND OF EFFECT.

FOR EXAMPLE, SEVERAL DIFFERENT HORMONES CAN INCREASE BLOOD SUGAR LEVELS (PAGE 221) INCLUDING GLUCAGON, ADRENALINE, AND GLUCOCORTICOID.

THEY ALL HAVE A SIMILAR EFFECT ON THE BLOOD SUGAR LEVEL, BUT THEY USE DIFFERENT MECHANISMS.

THEY COME FROM DIFFERENT SOURCES TOO—GLUCAGON IS SECRETED BY THE PANCREAS, WHILE ADRENALINE AND GLUCOCORTICOIDS ARE SECRETED BY THE ADRENAL GLANDS.

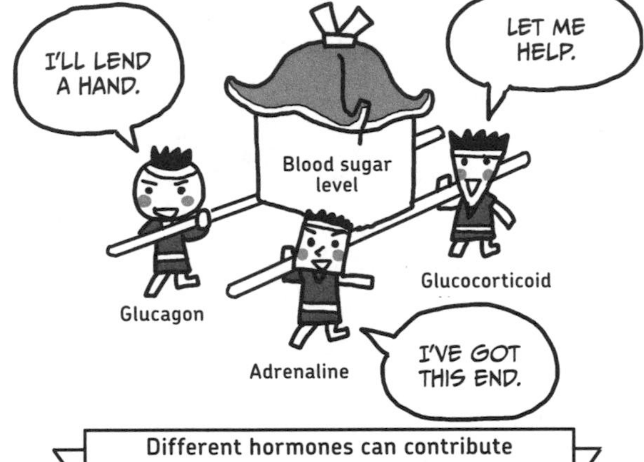

Different hormones can contribute to the same outcome.

BAM
P-PROFESSOR!
CAN I START RIGHT AWAY?
PLEAAAASSSEEE!
WHAT'S GOING ON HERE? YOUR OUTFIT...
IS THIS SOME KIND OF JOKE?
NO, NO, NOT AT ALL.
I'M RUNNING IN THE MARATHON TODAY.
AH...
I SUPPOSE YOU CAN BEGIN.
I HOPE YOU KEPT YOUR PRIORITIES STRAIGHT.
YOU HAVE YOUR WORK CUT OUT FOR YOU...
BETTER HURRY...
WHILE YOU CAN STILL REMEMBER SOMETHING FROM YOUR LAST-MINUTE CRAM SESSION.
SHEESH!
BLEHH
I WASN'T JUST MEMORIZING FACTS ANYWAY!

Osmotic Pressure
Concen-trated
Diluted
カリ
カリ
カリ
SCRIBBLE
SCRIBBLE
SCRIBBLE
TAP

I ALWAYS HAVE SUCH A HARD TIME WITH THAT AWFUL ENDOCRINE SYSTEM.
Endocrine System

HA HA HA
BUT YESTERDAY, I GOT IT ALL UNDER CONTROL, THANKS TO PROFESSOR KAISEI.

HURRAY!
HUH?
WHA-WHAT'S THAT?!
NO TALKING DURING THE EXAM!
UH-OH! THE MARATHON HAS ALREADY STARTED.
CLICK
CLICK
I CAN HEAR THE STARTER PISTOL...

BANG!
WHERE'S KUMIKO?
I WONDER WHY SHE'S NOT HERE YET!
HURRAH!
SHE MUST BE STUCK AT HER PHYSIOLOGY MAKEUP EXAM!
HURRAH!
NO WAY! THAT STINKS!
OK!
I FINISHED!!
HERE IT IS, PROFESSOR.
ZOOM
PLEASE GRADE THIS FOR ME!
WHOA!!
HUH, SHE REALLY DID FINISH IT.

EVEN MORE ABOUT THE ENDOCRINE SYSTEM!

Now that you have a general understanding of the endocrine system, I'll talk more about the main endocrine glands, one at a time. At the end of the chapter, you'll find a handy summary with the names and actions of the hormones secreted by each endocrine gland.

THE HYPOTHALAMUS AND PITUITARY GLAND

Let's start with the hypothalamus and pituitary gland. It's helpful to think of these as the headquarters or control center of the endocrine system. Many of the hormones released by the hypothalamus and pituitary gland act as signals to other endocrine glands, telling them to secrete their own hormones. The *hypothalamus*, which is located above the pituitary gland, interacts with both the nervous system and the endocrine system (Figure 10-1).

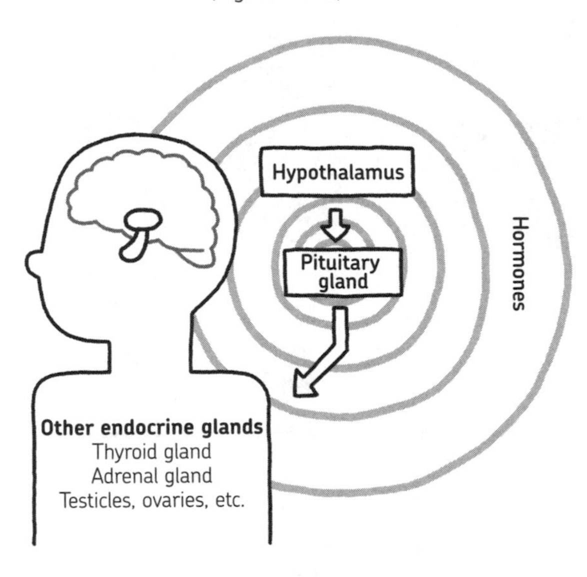

Figure 10-1: Hypothalamus and pituitary gland

The *pituitary gland* consists mainly of an anterior lobe and a posterior lobe. The anterior lobe produces and releases six hormones that stimulate other endocrine glands. These hormones, whose release is regulated by the hypothalamus, are a control mechanism for the endocrine system as a whole.

The posterior lobe of the pituitary gland secretes two types of hormones. However, the posterior pituitary does not produce these hormones. Instead, they are created by specialized neurons that stretch down from the hypothalamus, transporting molecules down their axons to the posterior pituitary, where they are released into the blood. In other words, the posterior pituitary is simply a release outlet (see Figure 10-2).

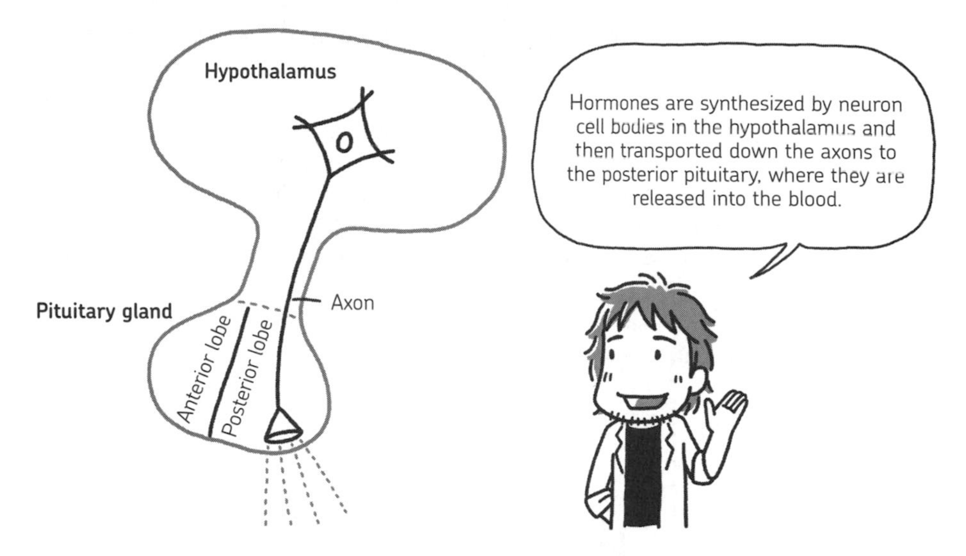

Figure 10-2: The hypothalamus sends hormones to the posterior pituitary.

> **Growth Hormone Disorders**
>
> One of the hormones produced by the pituitary gland is *growth hormone*. If too much growth hormone is produced during childhood, long bones such as those in the legs will continue to grow, causing a person to become extraordinarily tall. This condition is called *gigantism*. If growth hormone is produced excessively in an adult (due to a pituitary tumor, for example), then a person's hands, feet, and jaw become enlarged. This condition is called *acromegaly*. It's treated primarily with synthetic forms of somatostatin, a hormone that inhibits growth hormone.

THE THYROID AND PARATHYROID

The *thyroid gland* (or simply the *thyroid*) is located in the neck. It's regulated by the thyroid-stimulating hormone (TSH), which is secreted by the anterior pituitary.

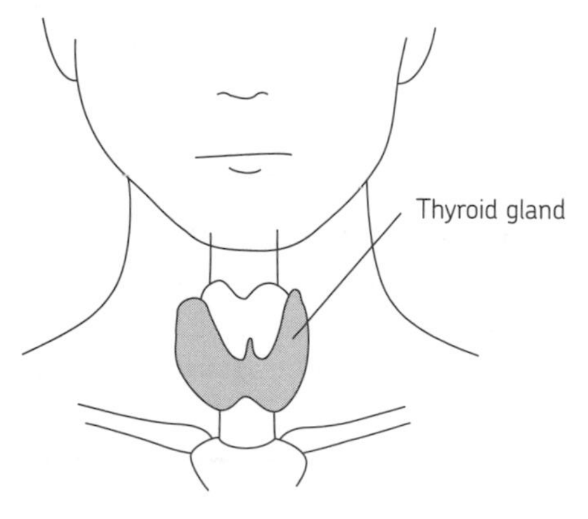

Figure 10-3: Location of the thyroid gland

The thyroid hormones include *thyroxine* (T4) and *triiodothyronine* (T3)—the numbers represent the quantity of iodine atoms per molecule in that hormone. These thyroid hormones accelerate *basal metabolism*—the amount of energy the body uses while at rest. If there's an excess of these hormones, the body will consume energy as if it were highly active even if it's at rest, potentially leading to fatigue. As shown in Figure 10-4, this may be accompanied by symptoms such as tachycardia (an abnormally fast resting heart rate), protruding eyeballs, and an enlarged thyroid. *Graves' disease* (also called *Basedow syndrome*) is a well-known type of hyperthyroidism that can produce such symptoms.

On the other hand, if the thyroid hormone level is too low, a person experiences a drop in metabolism, which in turn can cause listlessness, a decrease in body temperature, edema (swelling of body tissues), and a decrease in perspiration (see Figure 10-4).

Figure 10-4: Symptoms of an over- or underactive thyroid

Four small endocrine glands, called *parathyroid glands*, are attached to the thyroid gland. The parathyroid glands are named for their proximity to the thyroid, but they are completely independent and serve a different function than the thyroid. They secrete a *parathormone (PTH)*, which increases the blood calcium level (see Figure 10-5).

Overactive Parathyroid Glands

If the parathyroid glands are too active, too much bone will be broken down, making the bones brittle. This would also lead to *hypercalcemia*, or too much calcium in the blood, which can cause bone pain, weakness, and fatigue, and an increased risk of kidney stones.

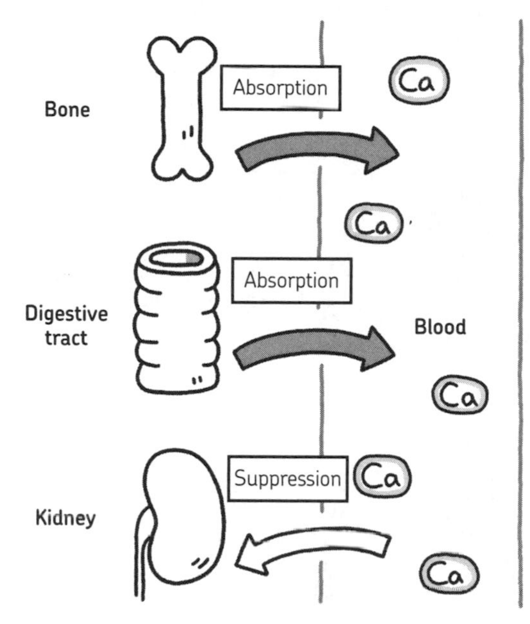

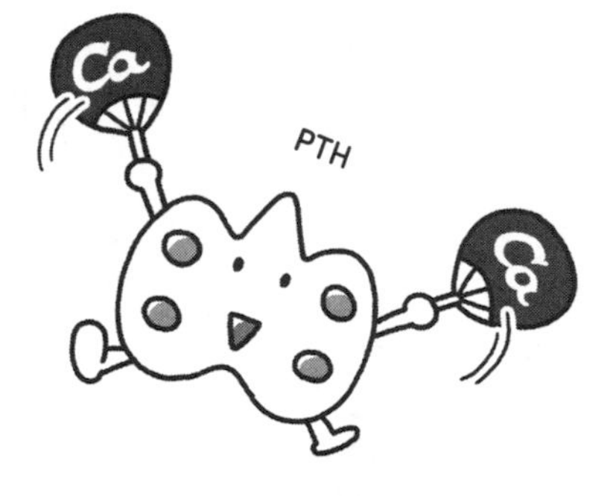

PTH promotes absorption of calcium from bones and the intestines, and suppresses excretion of calcium from urine, which increases the blood calcium level.

Figure 10-5: Functions of parathormone (PTH)

The blood calcium level is extremely important, isn't it?

It sure is. *Calcium* is indispensable to functions like muscle contraction, nerve transmission, and blood coagulation. If there is too little calcium in the blood, muscles can no longer move smoothly. Parathormones keep the blood calcium level from dropping too low.

THE ADRENAL GLANDS

The adrenal glands, situated on top of the kidneys, consist of an *adrenal cortex* and an *adrenal medulla*, each of which secretes different hormones (Figure 10-6).

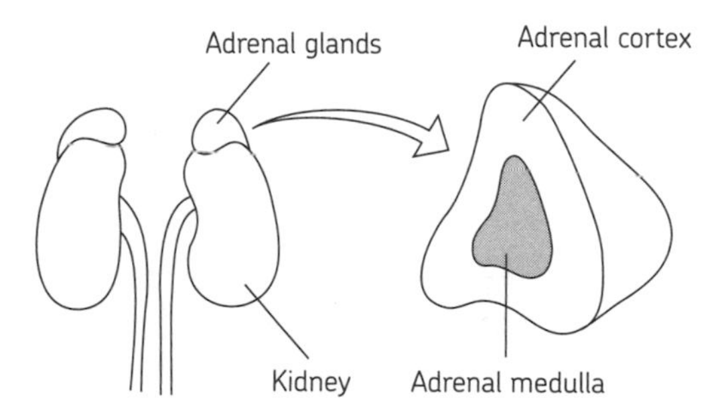

Figure 10-6: Adrenal cortex and adrenal medulla

The adrenal cortex secretes *steroid hormones*, meaning that they are synthesized from cholesterol. Although cholesterol has a bad reputation, it is a necessary component of the human body. The three types of hormones secreted by the adrenal cortex are glucocorticoids, mineralocorticoids, and androgens—each of these comes from a different layer of the cortex (Figure 10-7).

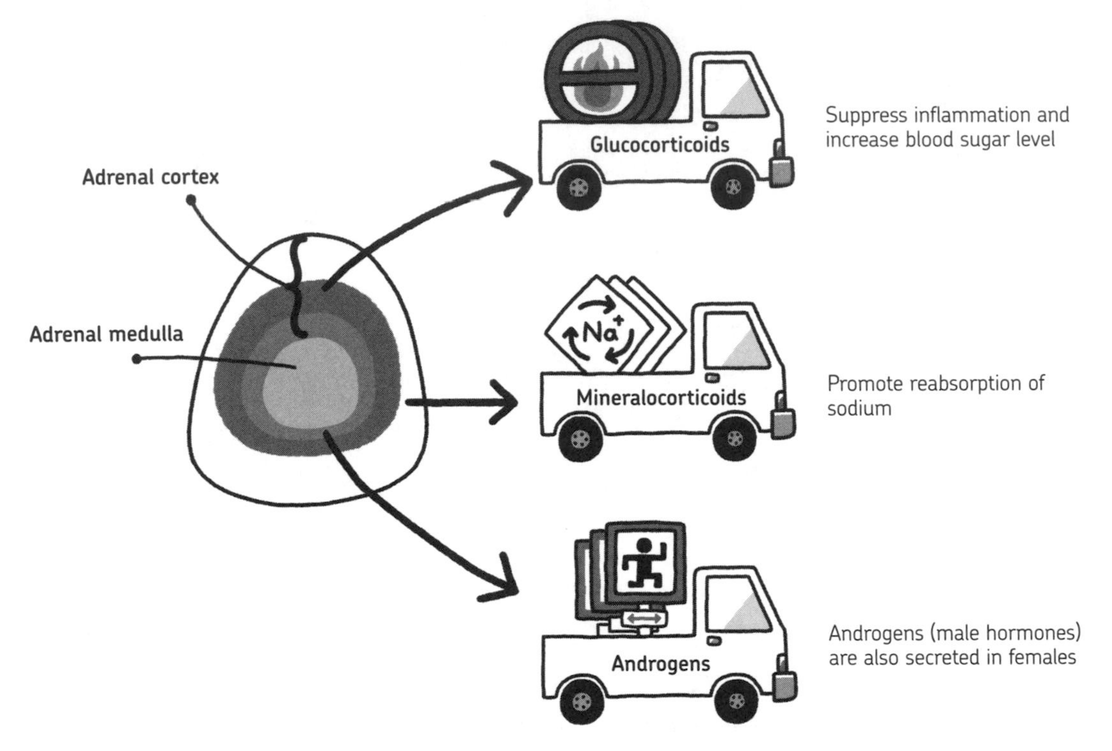

Figure 10-7: Functions of the adrenal cortex hormones

Glucocorticoids are involved in mobilizing glucose into the bloodstream, which is why *gluco* is part of their name. They also suppress inflammation and other immune system responses, and are widely used in pharmaceutical drugs.

Mineralocorticoids affect the balance of water and sodium in the body. The most important hormone in this category is aldosterone. *Aldosterone* influences the renal tubule of the kidneys to promote the reabsorption of sodium into the bloodstream. This draws water into the bloodstream as well and reduces the volume of urine, keeping more fluid in the body (see "Reabsorbing Water and Nutrients" on page 87).

Androgens are secreted by the adrenal cortex. Androgens are often called "male hormones," but these hormones are produced by the adrenal cortex in both males and females.

The adrenal cortex hormones are regulated by the *adrenocorticotropic hormone*, which is secreted by the anterior pituitary, right?

That's right. And if too many glucocorticoids are secreted by the adrenal cortex, the amount of adrenocorticotropic hormone is reduced to balance things out. This is an example of a negative feedback mechanism (as shown in "Balancing Hormone Levels" on page 207).

> **Androstenedione: A Sex Hormone Precursor**
>
> *Androstenedione* is an androgen secreted by the adrenal glands that is converted into testosterone and estrogen in fat and other tissues around the body. The level of estrogen produced by the adrenal glands is tiny compared to that produced by the ovaries in younger women, but this small amount is necessary in men and post-menopausal women.

Next, let's talk about the adrenal medulla, which secretes *adrenaline*. The adrenal medulla releases more adrenaline when it's triggered by the sympathetic nervous system (as part of what's often called a fight-or-flight response). In other words, adrenaline levels increase when you are excited, scared, or in the middle of strenuous activity. The adrenal medulla is almost like an extension of the sympathetic nervous system (see Figure 10-8).

Figure 10-8: The adrenal medulla and sympathetic nervous system work together to release high levels of adrenaline.

THE PANCREAS

The *pancreas* functions as both an exocrine gland and an endocrine gland. An *exocrine gland* secretes fluid by way of a duct to another area either inside or outside the body. In this case, the fluid is a digestive pancreatic juice secreted into the duodenum via the pancreatic duct. But as an *endocrine gland*, the pancreas also secretes hormones directly into the bloodstream. This endocrine function is performed by clusters of cells that are scattered throughout the pancreas like islands. Together they are called the *islets of Langerhans* (Figure 10-9). The islets of Langerhans include A cells (alpha cells) that secrete glucagon and B cells (beta cells) that secrete insulin.

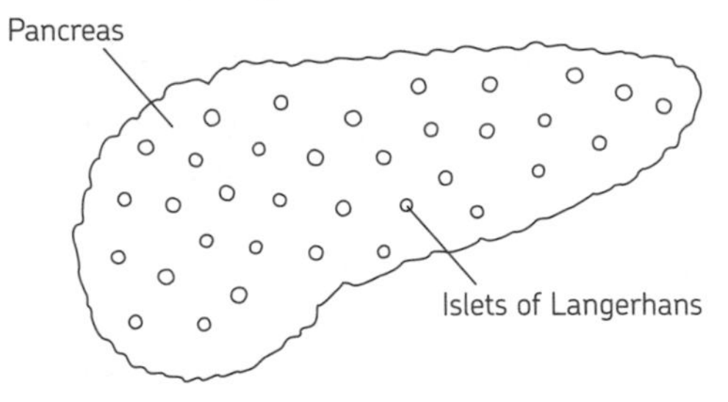

Figure 10-9: The pancreas contains more than a million islets of Langerhans.

Insulin regulates the body's blood sugar level. If the blood sugar level increases, the pancreas releases more insulin, which lowers the blood sugar level (Figure 10-10).

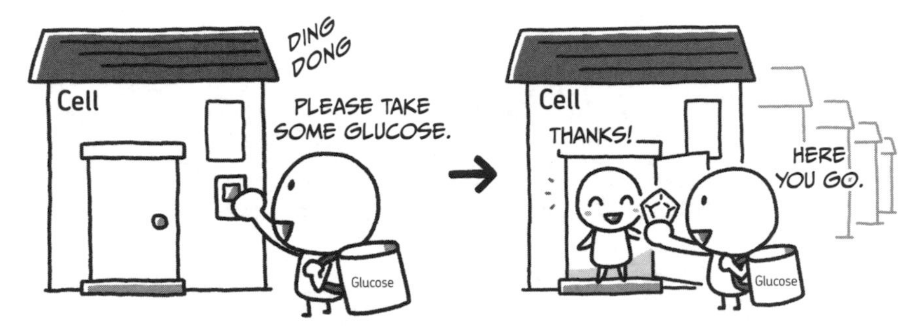

Figure 10-10: Insulin asks muscle and fat cells to absorb excess glucose to reduce sugar levels in the blood.

If I'm not mistaken, insulin is the only hormone that decreases the blood sugar level, right?

You are correct. Lots of hormones increase the blood sugar level (including adrenaline, growth hormones, glucocorticoid, and thyroid hormones), but insulin is the only one that lowers it (Figure 10-11). This is why insulin is so important.

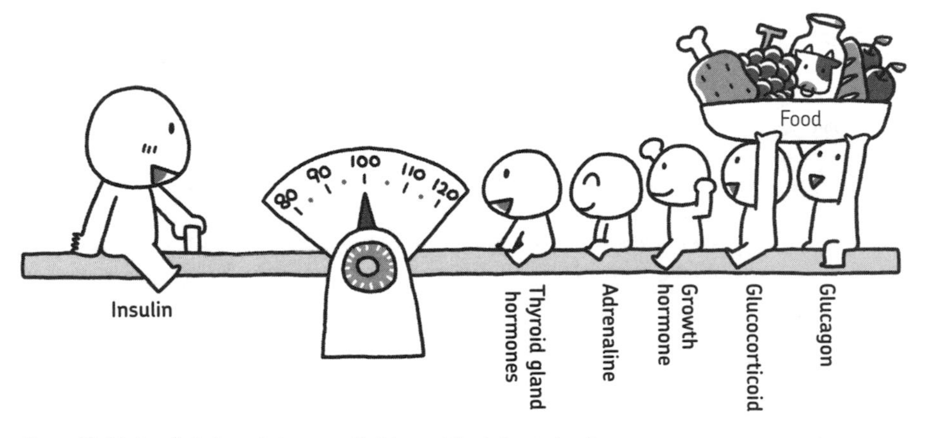

Figure 10-11: Insulin is the only hormone that lowers blood glucose levels.

If there's not enough insulin or if cells don't properly respond to the insulin, the blood sugar level will rise. This condition is called *diabetes mellitus*. People with this condition must carefully regulate their diet and may need to take medication or administer extra insulin by injection.

Diabetes Types 1 and 2

Diabetes mellitus is classified as either type 1 or type 2. Type 1 is caused by the loss of the ability to produce insulin. It usually first appears in childhood. While incurable, it can be managed by the administration of insulin injections several times a day. People with diabetes measure their blood sugar levels to know when to either eat something or inject insulin.

Type 2 diabetes occurs when cells stop responding well to insulin and therefore take less glucose from the bloodstream. This disease usually first appears later in life, and it's associated with lifestyle risk factors and other metabolic disorders like obesity. Treatments target different parts of the glucose-regulating system: sugar intake in the diet, the cells that have become less sensitive to insulin, the liver (which releases glucose), and the pancreas (which is still producing insulin).

Glucagon does the opposite of insulin: it raises the blood sugar level by breaking down glycogen in the liver, which releases glucose into the bloodstream (Figure 10-12).

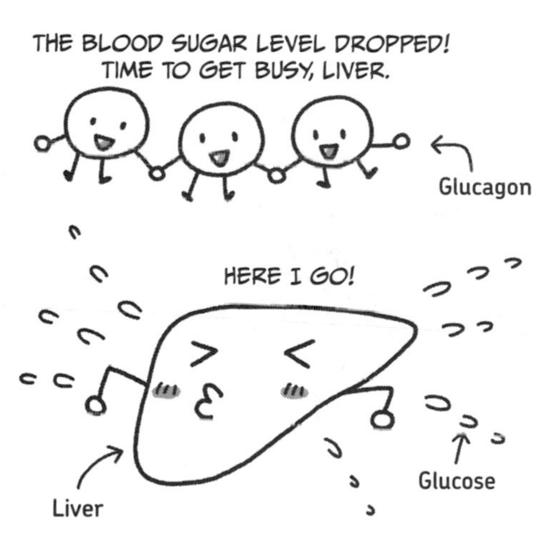

Figure 10-12: When blood sugar levels drop, glycogen is broken down in the liver to produce glucose and release it into the blood.

SEX HORMONES

Finally, we have the sex hormones, which are responsible for the development of primary and secondary sex characteristics. *Primary sex characteristics* are those that are already formed and recognizable at birth, such as genitals. *Secondary sex characteristics*, on the other hand, appear later in life, mostly starting with puberty. Male secondary sex characteristics include a deepened voice and accelerated growth of body and facial hair. In females, hormones trigger the development of breasts and menstruation.

Male hormones (also called *androgens*) are mainly secreted by the testicles, but smaller amounts are also secreted by the adrenal cortex. Female hormones include *estrogen* and *progesterone*. Both are secreted by the ovaries and are regulated by *gonadotropins* from the pituitary gland.

Periods are caused by estrogen and progesterone, right?

Yes, the menstrual cycle occurs as a result of female hormones, whose main function is to assist with conception and childbirth. The ovaries and uterus prepare themselves for the possibility of a pregnancy, starting over every cycle if an egg is not fertilized.

Let's talk about estrogen and progesterone in a little more detail. As shown in Figure 10-13, estrogen secreted by the ovary causes an ovum (egg) in the ovary to mature until it's ready to be released on its journey toward the uterus, otherwise known as *ovulation*. At the same time, estrogen causes the endometrium, the lining of the uterus, to thicken in preparation for a fertilized egg. In other words, estrogen works to enable conception.

After ovulation occurs, the ovarian follicle becomes the corpus luteum, which secretes progesterone to enrich the endometrium (Figure 10-13). This makes it easier for a fertilized egg to implant itself. If the egg isn't fertilized, the endometrium is no longer needed, and is broken down during menstruation.

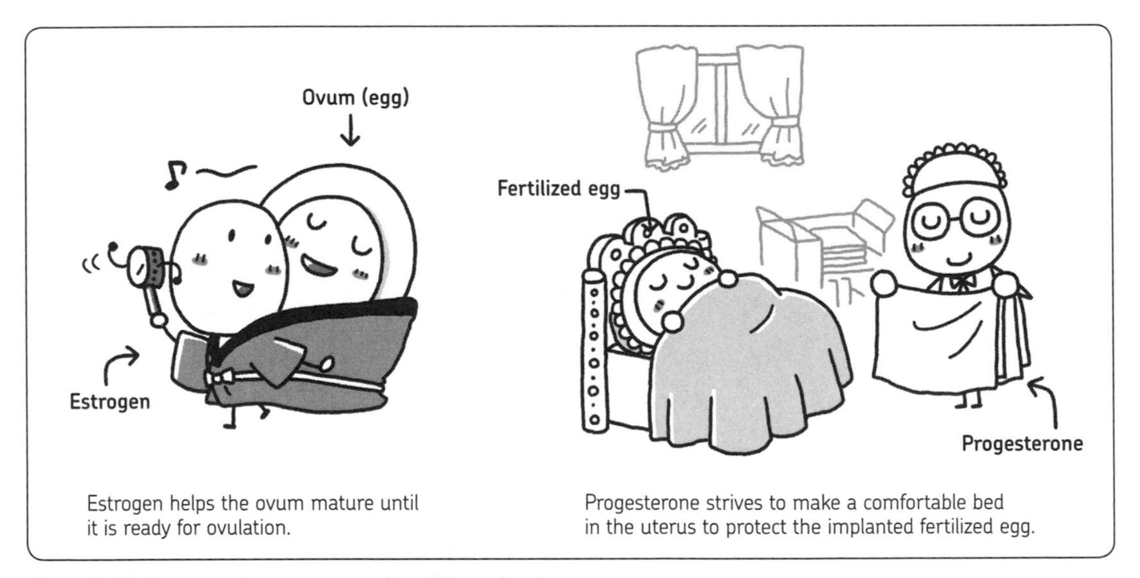

Figure 10-13: Estrogen and progesterone perform different functions.

We've gone over many different hormones. The diagram shown on page 221 provides a summary of the primary effects of the major hormones. Note that many have abbreviations; hormones are often referred to by their abbreviations in a clinical setting.

Main Endocrine Organs and Hormones

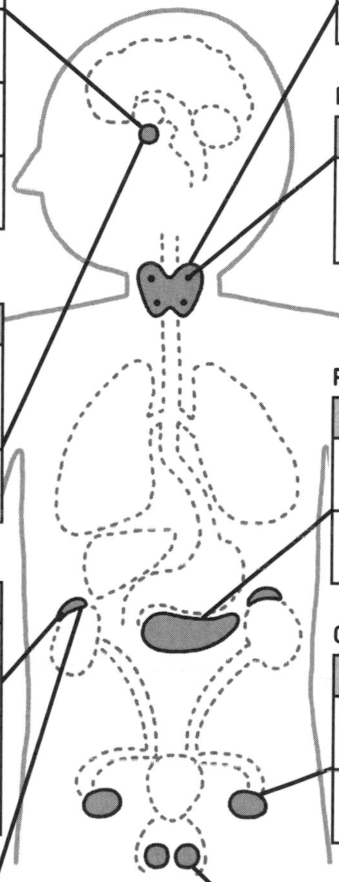

Anterior Pituitary Hormones

Hormone	Main effects
Growth hormone (GH)	Promotes bone growth
Thyroid-stimulating hormone (TSH)	Causes secretion of thyroid hormones
Adrenocorticotropic hormone (ACTH)	Causes secretion of adrenal cortex hormones
Follicle-stimulating hormone (FSH)	Promotes development of ovarian follicles
Luteinizing hormone (LH)	Causes the formation of the corpus luteum
Prolactin	Increases the production of milk

Posterior Pituitary Hormones

Hormone	Main effects
Vasopressin (antidiuretic hormone)	Promotes water reabsorption by the renal tubule of the kidneys
Oxytocin	Causes contraction of the uterus and lactation

Adrenal Cortex

Hormone	Main effects
Glucocorticoids	Inhibits inflammation and increases the blood sugar level
Mineralocorticoids	Promotes sodium (Na) reabsorption in the kidneys

Adrenal Medulla

Hormone	Main effects
Adrenaline	Increases blood pressure and stimulates the heart

Thyroid Gland

Hormone	Main effects
Triiodothyronine (T3)	Increases metabolism
Thyroxine (T4)	

Parathyroid Gland

Hormone	Main effects
Parathormone (PTH)	Increases the blood calcium level

Pancreas

Hormone	Main effects
Insulin	Decreases the blood sugar level
Glucagon	Increases the blood sugar level

Ovaries

Hormone	Main effects
Estrogen	Enables pregnancy to occur
Progesterone	Maintains pregnancy

Testicles

Hormone	Main effects
Androgens	Controls male characteristics

THWOP
SHE MADE IT! GO KUMIKO!
THEY'RE NOT SO FAR OFF...
SUU-HAH
IF I CAN KEEP IT UP, I CAN CATCH UP!
I CAN WIN THIS!
THROB
YOWCH
WELL, I WOULDN'T HAVE GUESSED...
SUU-HAH
SUU-HAH
I STUMBLED AGAIN.
OUCH!
THROB

I TRAINED HARD, AND I STUDIED EVEN HARDER.
OW
SUU-HAH
THERE'S NO WAY I'M GOING TO GIVE UP.
GOAL
SUU-HAH
THE FINISH LINE! FINALLY...
PLOP
YOU REALLY STUCK IT OUT.
PROFESSOR KAISEI...
I KEPT THINKING I COULDN'T GO ON...
BUT...YOU... PROFESSOR...
IT'S AMAZING... YOU PASSED.
AND WITH FLYING COLORS. IF YOU HAD SHOWN YOUR TRUE ABILITIES FROM THE START...

?
I PASSED THE MAKEUP EXAM?!
HEY!
NOOOOO!!! !!
YOU'RE NOT PROFESSOR KAISEI!
PUSH
ACK! I THOUGHT YOU'D WANT TO KNOW!
HURRAY!
YOU WERE AWESOME!
YOU WERE ABSOLUTELY THE BEST OF ALL THE RUNNERS TODAY!
YOU WERE SO GUTSY.
TH-THANK YOU.
?
HERE, TAKE THIS!
THIS GUY IS MY FAVORITE!
ME TOO... HERE'S MINE.
TAKE MINE TOO!
MINE TOO!
NO, TAKE MINE!
WHOA, GUYS!
JABBER
JABBER
HEY! GIVE HER SOME SPACE...

WELL, MS. KARADA, YOU'RE QUITE THE INSPIRATION.
MY KIDS HAVE TAKEN A REAL SHINE TO YOU.
YOUR KIDS...?
SOLEMN NOD
DADDY!
WON'T YOU CONSIDER TRANSFERRING TO THE DEPARTMENT OF SPORTS AND HEALTH SCIENCE?
YOU'RE A NATURAL ATHLETE AND A LEADER, TOO.
AND THERE'S SO MUCH MORE TO LEARN ABOUT PHYSIOLOGY!
WELL...
I KNOW I HAVE LOTS LEFT TO LEARN FROM YOU!

...
HA
HA
I'LL DEFINITELY THINK ABOUT IT.

* FROM LEFT TO RIGHT: *KAGERO* = MAYFLY; *AGEHA(CHOU)* = SWALLOWTAIL BUTTERFLY; *HOTARU* = FIREFLY; *AKIAKANE* = RED DRAGONFLY; *HACHI* = BEE

AFTERWORD: CREATING THIS BOOK

What's the best way to learn about physiology? That's what we asked ourselves when writing this book, and I hope that Osamu and Kumiko have shown that the best way to learn physiology is not through rote memorization alone.

Kumiko soon finds that physiology is so much more interesting when you take a personal interest and identify with the processes you're learning about. Also, while physiology certainly requires some memorization, it's important to have a wider understanding of how the different parts of our bodies work as a whole. Each organ has its own functions, but the organs also act in conjunction with each other. Similarly, blood, oxygen, nerves, hormones, and lymph fluid work *together* to carry out various functions in a huge network. Reading this book will help you understand these relationships, and so better understand the human body.

We were aware from the start that covering such an extensive academic subject in a single book would be a formidable endeavor. For that reason, we used memorable illustrations and scenes, instead of difficult diagrams and anatomical charts, to help readers see physiology as an interesting field and to win over those who previously disliked the subject.

This manga shouldn't be the only text you work from when studying for your physiology exam, but we'd be extremely pleased if it were to give you an interest and basic education in physiology and encourage you to keep studying.

Last but not least, we would like to take this opportunity to express our sincere thanks to Professor Etsuro Tanaka of Tokyo University of Agriculture, who provided editorial supervision in all details of this book; Ms. Yasuko Suzuki, a medical writer who collaborated; and everyone in the Development Department at Ohmsha, Ltd.

BECOM CO., LTD.
OCTOBER 2011

INDEX

M

N

O

S

T

U

V

W

Z

ABOUT THE AUTHOR

Etsuro Tanaka is a doctor of medicine who specializes in physiology and nutritional science. A professor in the Tokyo University of Agriculture Faculty of Applied Bio-Science, he has also written several popular physiology textbooks for nursing students.

PRODUCTION TEAM FOR THE JAPANESE EDITION

Production: BeCom Co., Ltd.

Since its foundation in 1998 as an editorial and design production studio, BeCom has produced many books and magazines in the fields of medicine, education, and communication. In 2001, BeCom established a team of comic designers, and the company has been actively involved in many projects such as manga books, corporate guides, and product manuals. For more information about BeCom, visit *http://www.becom.jp/*.

Yurin Bldg 5F, 2-40-7 Kanda-Jinbocho, Chiyoda-ku, Tokyo, Japan 101-0051

Tel: 03-3262-1161; Fax: 03-3262-1162

Drawing: Keiko Koyama (Koguma Workshop; *http://www.koguma.info*)

Text illustration: Bazzy

Writing collaboration: Yasuko Suzuki

Scenario: Tomohiko Tsuge and Eiji Shimada (BeCom)

Cover design: Keiji Ogiwara (BeCom)

DTP and editing: BeCom Co., Ltd.

HOW THIS BOOK WAS MADE

The *Manga Guide* series is a co-publication of No Starch Press and Ohmsha, Ltd. of Tokyo, Japan, one of Japan's oldest and most respected scientific and technical book publishers. Each title in the best-selling *Manga Guide* series is the product of the combined work of a manga illustrator, scenario writer, and expert scientist or mathematician. Once each title is translated into English, we rewrite and edit the translation as necessary and have an expert review each volume. The result is the English version you hold in your hands.

COLOPHON

The Manga Guide to Physiology is set in CCMeanwhile and Chevin. The book was printed and bound by Edwards Brothers Malloy in Ann Arbor, Michigan. The paper is 60# Husky Offset Smooth, which is certified to the Sustainable Forestry Initiative® (SFI®) standard.

MORE MANGA GUIDES

Find more *Manga Guides* at your favorite bookstore, and learn more about the series at *http://www.nostarch.com/manga*.